Tracking with Particle Filter for High-dimensional Observation and State Spaces

Series Editor
Henri Maître

Tracking with Particle Filter for High-dimensional Observation and State Spaces

Séverine Dubuisson

iSTE

WILEY

First published 2015 in Great Britain and the United States by ISTE Ltd and John Wiley & Sons, Inc.

ISTE Ltd
27-37 St George's Road
London SW19 4EU
UK

www.iste.co.uk

John Wiley & Sons, Inc.
111 River Street
Hoboken, NJ 07030
USA

www.wiley.com

Library of Congress Control Number: 2014955871

British Library Cataloguing-in-Publication Data
A CIP record for this book is available from the British Library
ISBN 978-1-84821-603-7

Contents

NOTATIONS ix

INTRODUCTION xi

CHAPTER 1. VISUAL TRACKING BY PARTICLE FILTERING 1

1.1. Introduction 1
1.2. Theoretical models 2
1.2.1. Recursive Bayesian filtering 2
1.2.2. Sequential Monte-Carlo methods 4
1.2.3. Application to visual tracking 8
1.3. Limits and challenges 18
1.4. Scientific position 22
1.5. Managing large sizes in particle filtering 22
1.6. Conclusion 26

CHAPTER 2. DATA REPRESENTATION MODELS . . . 29

2.1. Introduction 29
2.2. Computation of the likelihood function 30
2.2.1. Exploitation of the spatial redundancy 31
2.2.2. Exploitation of the temporal redundancy . . . 42

2.3. Representation of complex information 50
2.3.1. Representation of observations for movement detection, appearances and disappearances 50
2.3.2. Representation of deformations 53
2.3.3. Multifeature representation 56
2.4. Conclusion 75

CHAPTER 3. TRACKING MODELS THAT FOCUS ON THE STATE SPACE 79

3.1. Introduction 79
3.2. Data association methods for multi-object tracking 80
3.2.1. Particle filter with adaptive classification 84
3.2.2. Energetic filter for data association 87
3.3. Introducing fuzzy information into the particle filter 95
3.3.1. Fuzzy representation 96
3.3.2. Fuzzy spatial relations 98
3.3.3. Integration of fuzzy spatial relations into the particle filter 99
3.4. Conjoint estimation of dynamic and static parameters 114
3.5. Conclusion 119

CHAPTER 4. MODELS OF TRACKING BY DECOMPOSITION OF THE STATE SPACE 123

4.1. Introduction 123
4.2. Ranked partitioned sampling 126
4.3. Weighted partitioning with permutation of sub-particles 133
4.3.1. Permutation of sub-samples 135
4.3.2. Decrease the number of resamplings 138
4.3.3. General algorithm and results 138
4.4. Combinatorial resampling 142
4.5. Conclusion 149

CHAPTER 5. RESEARCH PERSPECTIVES IN TRACKING AND MANAGING LARGE SPACES 151

5.1. Tracking for behavioral analysis: toward finer tracking of the "future" and the "now" 153

5.2. Tracking for event detection: toward a top-down model 156

5.3. Tracking to measure social interactions 159

BIBLIOGRAPHY 163

INDEX 197

Notations

$\mathbb{P}(.)$	Probability Distribution
$p(.)$	Probability density
$t = 1, \ldots, T$	Time indices
$i = 1, \ldots, N$	Particle indices
$m = 1, \ldots, M$	State vector component indices
$l = 1, \ldots, L$	Observation vector component indices
$\mathcal{I}_t$	Image at instant t
$\mathcal{I}_t(x, y)$	Intensity of pixel (x, y) of the image at instant t
W, H	Width and height of an image or a region
$\mathcal{X} = \mathcal{X}^1 \times \ldots \times \mathcal{X}^M$	State space, decomposable as a product of sub-spaces $\mathcal{X}^m$
$\lvert\mathcal{X}\rvert$	Dimension of the state space
$\mathcal{Y} = \mathcal{Y}^1 \times \ldots \times \mathcal{Y}^L$	Observation space, decomposable as a product of sub-spaces $\mathcal{Y}^l$
$\mathbf{x}_t \in \mathcal{X}$	State vector at instant t
$\mathbf{x}_t^{(i)} \in \mathcal{X}$	Particle i at instant t
$\mathbf{x}_t^m \in \mathcal{X}^m$	Component (or part) m of state vector at instant t (a feature, a sub-region, an object in the multi-object case, *etc.*)

$\mathbf{x}_t^{m,(i)} \in \mathcal{X}^m$	Component (or part) m of particle i at instant t
$\mathbf{y}_t \in \mathcal{Y}$	Observation vector at instant t
$\mathbf{y}_t^l \in \mathcal{Y}^l$	Component (or measurement) l of observation vector at instant t (a view, a modality, *etc.*)
$\mathcal{S}_t = \{\mathbf{x}_t^{(i)}, w_t^{(i)}\}_{i=1}^N$	Weighted sample, or set of N particles, at instant t

Introduction

With the progress made in the domains of electronics and microelectronics, the acquisition of video sequences became a task of particular triviality. Hence, in computer vision, algorithms working with video sequences have undergone considerable development over the past few years [SZE 10]. Skimming through a book dedicated to computer vision, written 30 years ago [BAL 82], we note that the notion of movement is discussed nearly in terms of approximation: here, the issue was detecting movement, rather than analyzing it. In particular, analysis of the optical flow [BAR 94], very popular at the time, only allowed characterizing temporal changes within the sequence. Little by little, with the rapid improvement of sensor quality and therefore of the resolution of the images they provided, as well as computer processing power and memory, it became possible, perhaps essential, to analyze movement in addition to detecting it: where does it come from? What behavior does it reflect? Hence, new algorithms made their appearance [SHI 94], their purpose being to detect and to follow entities in a video sequence. These are grouped under the name of tracking algorithms. Today, tracking single and multiple objects in video sequences is one of the major themes of computer vision. There are in fact many practical applications, notably in human–machine interaction, augmented reality, traffic control, surveillance, medical or biomedical imagery and even interactive games. The diversity of the problems to solve, as well as computing challenges created by object tracking in video sequences, motivates an increasing amount of research every year. Thus, in the year 2012 alone, in three major computer vision conferences (IEEE Conference on Computer Vision and Pattern Recognition (CVPR), European Conference on Computer Vision (ECCV) and British Machine Vision Conference (BMVC)), three workshops and two tutorials were dedicated to tracking; we can even mention the PETS workshop (International Workshop on Performance Evaluation of Tracking and Surveillance), which suggests a competition around tracking every two years. Furthermore, databases are increasingly available to allow researchers to compare their results [SIG 10a, WU 13][1]. The intense research activity around object tracking in video sequences is explained by the large amount of challenges that it undertakes. Indeed, it requires efficiently extracting the information related to the object or objects to track from the images, modeling it to obtain a representation that is both precise and compact and solving the compromise between tracking quality and efficiency. It is then required to be able to create temporal links between the object instances in each time step, while managing the occasional appearances and disappearances of objects on the scene. Finally, it is sometimes necessary to extract meta-data to respond to the needs of a specific application (behavioral analysis, detection of an event, etc.). In addition to these difficulties, there are those induced by the state of the object (appearance and deformation), the variations of the illumination of the scene, the noise present in the images, object occlusion, etc. Hence, object tracking reveals itself as a very complex process, especially given the ever-growing requirements in terms of

1 A non-exhaustive list is available at http://clickdamage.com/sourcecode/cv_datasets.php.

tracking quality and processing speed in practical applications.

Over the last few years, sequential Monte Carlo methods [DOU 01, GOR 93, ISA 98a], better known as particle filters, became the algorithm for visual tracking *par excellence*. Their aim is to estimate the density of the filtering that links the states of the tracked objects to previous and current observations by approximating it using a weighted sample. Outside of the simplicity of their implementation, these approaches are capable of maintaining multiple hypotheses over time, which makes them robust to the challenges of visual tracking. Additionally, given their probabilistic nature, their very generic formalism makes it possible to consider complex modeling for the available objects and observations, whose densities could be non-parametric and/or multimodal. Nevertheless, their use requires making sure to stay within their mathematical framework, which is rigorous in spite of its simplicity. Moreover, we need to make sure that algorithmic costs remain reasonable (by incorporating, for example, independence hypotheses when they are justifiable). We positioned ourselves naturally in this methodological context, particularly by noting that some of the primary advantages offered by particle filtering cannot, at the current time, be used without making a certain number of often simplifying hypotheses. Specifically, if maintaining multiple hypotheses over time presents a real advantage to particle filtering, the minimal amount to maintain a good approximation of the filtering density is as high as the chosen data model leads to a high-volume representation. This results in serious problems once we attempt to refine the representation by integrating, for example, all the richness of information that is supplied by the sensors.

The goal of this book is to present the various contributions related to managing large state and observation representation spaces, which we consider to be one of the major challenges of particle filtering today. We distinguished three primary axis that guided the research and will be the subject of Chapters 2 through 4.

The first axis concerns the choice of the data model in order to lighten the representation, as well as accelerate its extraction. The work on this axis is essential to simplifying the calculations related to the estimation by particle filtering. Indeed, in order to be solved in a robust and targeted manner, current tracking problems require exploiting a multitude of available informations/observations, whose quality is constantly improving. Therefore, this requires increasingly finer descriptions of the dynamic scene that is being processed, which have the tendency to put considerable weight on calculations. Today, however, there are reliable and efficient techniques for data extraction that allow better exploitation of image information, they are not necessarily appropriate to the case of particle filtering and its multihypothesis representation, as they lead to repeating calculations, which may be disastrous for the efficiency of the filter. Hence, histograms are a widely used representation model, however, their extraction can quickly become a bottleneck for the filter response time, so it is appropriate to find methods that are appropriate for their extraction. The size of the spaces that we are working on has a significant influence on the response times, and being able to combine a set of characteristics, observations, information and to supply a model described in less space is an equally essential task, undertaken by numerous researchers. As we will see later, efficient combinations can be made either during the correction precess, which can be considered as *a posteriori* fusion, or be integrated directly to the tracking process (propagation and correction).

Several suggested theoretical and/or algorithmic solutions were able to obtain either a significant decrease in the processing time of likelihood functions, notably with optimized histogram extractions, or original models of classical tracking problems, such as deformation of the object over time, its multiple possible representations (appearances, modalities and fragments) and even the detection of objects (new or known) between two images.

The second axis concerns the exploration of the state space. Indeed, the particle filter features a hypothesis (particle) propagation phase, aiming to diffuse hypothesis toward areas with high likelihood, where they will be attributed a high weight during the first correction stage. When the state space is large, detecting areas with high likelihood turns out to be difficult, as exhaustive exploration of the space requires multiplying the number of hypotheses, which is impossible within reasonable processing times. This problem can be solved in two different ways. The first solution constitutes a primary research axis and consists of "choosing" the areas of the space to explore, which we call "focusing". Hence, we can maintain a reasonable number of hypothesis, while smartly exploring the areas in which they will be propagated, areas with assumed high likelihood. This can be done either through detection before propagation (*detect-before-track*), or by defining dedicated proposition functions. It is this latter solution that we chose to develop in this book and that consists of decomposing the state space into subspaces of smaller sizes, that can be processed with "few" hypothesis.

We suggested several approaches that allow better focusing within the state space and, therefore, accelerate the process of tracking by particle filtering process. Two types of contributions dedicated to multiobject tracking allowed us not to have to process all the possible combinations associating measures with objects. The first type aims to, first of all, model a proposition function allowing the propagation of particles only in areas where movement was detected in advance, and which are seen as measurements, and then to classify these particles by associating them with objects on the scene. The second type takes into account the past dynamics of the objects and suggests a data association model that is very unconstrained, as it depends on few parameters. These models allow simply calculating the association probabilities between the measurements and objects. We will show that it is equally possible to input fuzzy spatial information into the particle filter. This allowed modeling of a new proposition function taking into account not only the current observation, but also the history of fuzzy spatial relations that characterized the past trajectories of the objects. The result allows considerably more flexible and better adapted tracking of sudden changes in trajectory or form. The appeal of this type of modeling is shown through various applications, such as object tracking, by managing erratic movements, multiobject tracking, by taking into account the case of object occlusion and finally multiform tracking, by taking into account deformable objects.

As the previous axis, the third one aims to make the exploration of large state spaces possible in practice. However, here we are no longer looking to work with the entire space whose exploration hyper-volume we reduced. Rather, we suggest decomposing it into subspaces of smaller sizes, in which calculations can be made within reasonable times, as they allow estimating distributions with fewer parameters than those required in the whole space. Hence, the latter is defined as a joint space. In this book, we are interested in “non approximated” decomposition methods, that is methods that guarantee asymptotically that the particles sample correctly the filtering distribution over the whole space. Thus, these methods do not make any simplifying hypothesis and only exploit the independences existing in the tracking problem. Among these techniques, partitioned sampling is used a lot today, although it has several limits. Indeed, this technique sequentially covers every subspace in order to construct, progressively, all the particles on the joint space (whole space), which can create problems if the order that the subspaces are processed in is completely arbitrary: if the sub-hypothesis made in the first subspace is incorrect, it will contribute to diminish the global score of the hypothesis as a whole, even if other sub-hypotheses are correct. Thus, the quality of the tracking will be low.

There are contributions that allowed us to solve this problem. First is the possibility to add the order in which objects need to be processed to the estimation process. This order is estimated sequentially, at the same time as the states of the objects, which allows taking the least reliable objects into account for tracking last. We can also exploit the conditional independences intrinsic to the problem of tracking (without making abusive hypothesis). This naturally leads to using dynamic Bayesian networks, rather than Markov chains, in order to model the filtering process. The exploitation of independence properties of these networks allowed us to develop a new method for permuting certain subsamples of subparticles that allow better estimations of the filtering density models, while guaranteeing that the estimated density remains unchanged. This method leads to reducing not only tracking errors, but also calculation times. This idea of permutation is exploited to suggest a new resampling method, which also allows us to significantly improve tracking.

The structure of this book is as follows. In Chapter 1, we will present the theoretical elements that are necessary to understand particle filtering. We will then explain how this methodology is used in the context of visual tracking, particularly the fundamental points to consider. This will eventually allow us to describe several current limits and challenges of tracking by particle filtering and, thus, justify our scientific position. Chapter 2 presents contributions related to modeling and extracting the data to process, as well as the choice for its representation to simplify, and thereby accelerating the calculations. In Chapter 3, we describe several contributions that allow exploring the state space by focusing on certain specific areas, considered more interesting than others. Chapter 4 shows, through certain works, how to decompose the state space into subspaces in which calculations are possible. Finally, we suggest a conclusion and an opening on the future of tracking, specifically by particle filtering, in Chapter 5.

1

Visual Tracking by Particle Filtering

1.1. Introduction

The aim of this introductory chapter is to give a brief overview of the progress made over the last 20 years in visual tracking by particle filtering. To begin (section 1.2), we will present the theoretical elements necessary for understanding particle filtering. Thus, we will first introduce recursive Bayesian filtering, before giving the outline of particle filtering. For more details, in particular theorem demonstrations and convergence studies, we invite the reader to refer to more advanced studies [CHE 03b, DOU 00b, GOR 93]. We will then explain how particle filtering is used in visual tracking in video sequences. Although the literature is abundant on this subject and evolving very fast, it is impossible to give a complete overview of this subject. Next, section 1.3 presents certain limits of particle filtering. Toward the end, we specify our scientific position in section 1.4 and the methodological axes that allow a part of these problems to be solved. Finally, section 1.5 gives the current state of the main large families of approaches that are concerned with managing large-sized state and/or observation spaces in particle filtering.

1.2. Theoretical models

1.2.1. *Recursive Bayesian filtering*

Recursive Bayesian filtering [JAZ 70] aims to approximate the state of a hidden Markov process, which is observed through an observation equation. Let $\{\mathbf{x}_{0:t}\} = \{\mathbf{x}_0, \ldots, \mathbf{x}_t\}$ be this process, where $\mathbf{x}_t$ is the state vector, $\mathbf{y}_t$ the observation at instant t and the two models:

$$\begin{cases} \mathbf{x}_t = f_t(\mathbf{x}_{t-1}, \mathbf{u}_t) \\ \mathbf{y}_t = g_t(\mathbf{x}_t, \mathbf{v}_t) \end{cases} \quad [1.1]$$

The first equation is the state equation, with the state transition function f_t between the instants $t-1$ and t, and the second is the observation equation, giving the measurement of the state through an observation function g_t. $\mathbf{u}_t$ and $\mathbf{v}_t$ are independent white noises.

All information necessary for approximating $\mathbf{x}_{0:t}$ is contained in the *a posteriori* density, also known as the filtering density, $p(\mathbf{x}_{0:t}|\mathbf{y}_{1:t})$, where $\mathbf{y}_{1:t} = \{\mathbf{y}_1, \mathbf{y}_2, \ldots, \mathbf{y}_t\}$, in which we can prove, by applying the definition of conditional probabilities, that it follows the following recursive equation for a known [CHE 03b] $t \geq 1$ ($p(x_0)$:

$$p(\mathbf{x}_{0:t}|\mathbf{y}_{1:t}) = \frac{p(\mathbf{y}_t|\mathbf{y}_{1:t-1}, \mathbf{x}_{0:t})p(\mathbf{x}_t|\mathbf{x}_{0:t-1}, \mathbf{y}_{1:t-1})p(\mathbf{x}_{0:t-1}|\mathbf{y}_{1:t-1})}{\int_{\mathbf{x}_{0:t}} p(\mathbf{y}_t|\mathbf{y}_{1:t-1}, \mathbf{x}_{0:t})p(\mathbf{x}_{0:t}|\mathbf{y}_{1:t-1})d\mathbf{x}_{0:t}} \quad [1.2]$$

Under the Markov hypothesis, $p(\mathbf{y}_t|\mathbf{y}_{1:t-1}, \mathbf{x}_{0:t}) = p(\mathbf{y}_t|\mathbf{x}_t)$ (the observations at different instants are independent between themselves given the states and do not depend on the state at the current instant) and

$p(\mathbf{x}_t|\mathbf{x}_{0:t-1}, \mathbf{y}_{1:t-1}) = p(\mathbf{x}_t|\mathbf{x}_{t-1})$ (the current state only depends on the previous state), equation [1.2] becomes:

$$p(\mathbf{x}_{0:t}|\mathbf{y}_{1:t}) = \frac{p(\mathbf{y}_t|\mathbf{x}_t)p(\mathbf{x}_t|\mathbf{x}_{t-1})p(\mathbf{x}_{0:t-1}|\mathbf{y}_{1:t-1})}{\int_{\mathbf{x}_{0:t}} p(\mathbf{y}_t|\mathbf{x}_t)p(\mathbf{x}_{0:t}|\mathbf{y}_{1:t-1})d\mathbf{x}_{0:t}} \qquad [1.3]$$

The state transition equation is represented by the density $p(\mathbf{x}_t|\mathbf{x}_{t-1})$ and is linked to the f_t function. This density is also called the transition function and gives the probably state $\mathbf{x}_t$ at the instant t, given its previous state $\mathbf{x}_{t-1}$. The observation equation is represented by $p(\mathbf{y}_t|\mathbf{x}_t)$ and is linked to the function g_t. This density is also called the likelihood function and gives the probability of making the observation $\mathbf{y}_t$ given the state $\mathbf{x}_t$. We can see that equation [1.3] is recursive and it decomposes into two primary stages that we detail below.

1) The first stage, known as prediction step, allows approximating the *a posteriori* density $p(\mathbf{x}_{0:t}|\mathbf{y}_{1:t-1})$ using the transition distribution $p(\mathbf{x}_t|\mathbf{x}_{t-1})$ and the previously approximated density $p(\mathbf{x}_{0:t-1}|\mathbf{y}_{1:t-1})$.

2) The second stage, called correction step, allows obtaining the *a posteriori* density $p(\mathbf{x}_{0:t}|\mathbf{y}_{1:t})$, using the likelihood distribution $p(\mathbf{y}_t|\mathbf{x}_t)$, which depends on the new observation. This *a posteriori* density represents the density of the probability to have the set of states $\mathbf{x}_{0:t}$, among all the possible states, given the history of the observations $\mathbf{y}_{1:t}$.

In order to obtain calculable estimators of $\mathbf{x}_{0:t}$, we can use, for example, the conditional mean, given by:

$$\mathbb{E}_p[\mathcal{F}(\mathbf{x}_{0:t})] = \int_{\mathbf{x}_{0:t}} \mathcal{F}(\mathbf{x}_{0:t})p(\mathbf{x}_{0:t}|\mathbf{y}_{1:t})d\mathbf{x}_{0:t} \qquad [1.4]$$

where $\mathcal{F}$ is some bounded function. If the densities are Gaussian, then there exists a solution (analytical expression of the Gaussian parameters to approximate) given by the Kalman filter [KAL 60]. Otherwise, the whole of equation [1.4] is not calculable directly. We can invoke, under special conditions, the solutions given by the following types of methods:

– analytical (extended Kalman filter [JAZ 70], unscented Kalman filter [JUL 97]) that approach the law from a Gaussian sum and are better adapted to weakly nonlinear and unimodal cases, which is nonetheless not appropriate for most problems of vision;

– numerical (approximations by discrete tables, division into parts) that are, most of the time, complex to solve, not very flexible and only adapted to state spaces of a small size.

Most of the time, in vision, solutions are not adapted as the integrals are not directly calculable. For the general case (non-parametric and multi-modal densities), it is necessary to make use of numerical approximations, such as those provided by sequential Monte-Carlo methods, which we will present in the following section and that are the methodological heart of this work.

1.2.2. *Sequential Monte-Carlo methods*

Sequential Monte-Carlo methods, also known under the name of particle filters (PFs), were studied by many researchers at the beginning of the 1990s [GOR 93, MOR 95] and combine Monte-Carlo simulation and recursive Bayesian filtering. Today, they are widely used in the computer visualization community. Before detailing the principle of particle filtering, we need to introduce importance sampling.

1.2.2.1. *Importance sampling*

Once the *a posteriori* density defined by equation [1.3] has been approximated, we can evaluate the estimator given in equation [1.4]. The Monte-Carlo method allows us to approximate this integral with the realization of a random variable distributed according to the *a posteriori* density.

Unfortunately, we are almost never able to sample following this law, so to solve this problem, we introduce a proposal function (or importance function) $q(\mathbf{x}_{0:t}|\mathbf{y}_{1:t})$, whose support contains $p(\mathbf{x}_{0:t}|\mathbf{y}_{1:t})$ and from which we can sample. The conditional mean is then given by:

$$\begin{aligned}\mathbb{E}_p[\mathcal{F}(\mathbf{x}_{0:t})] &= \int_{\mathbf{x}_{0:t}} \mathcal{F}(\mathbf{x}_{0:t})\frac{p(\mathbf{x}_{0:t}|\mathbf{y}_{1:t})}{q(\mathbf{x}_{0:t}|\mathbf{y}_{1:t})} q(\mathbf{x}_{0:t}|\mathbf{y}_{1:t})d\mathbf{x}_{0:t} \\ &= \mathbb{E}_q\left[\mathcal{F}(\mathbf{x}_{0:t})\frac{p(\mathbf{x}_{0:t}|\mathbf{y}_{1:t})}{q(\mathbf{x}_{0:t}|\mathbf{y}_{1:t})}\right]\end{aligned} \quad [1.5]$$

With N realizations $\mathbf{x}_{0:t}^{(i)} \sim q(\mathbf{x}_{0:t}|\mathbf{y}_{1:t})$, $i = 1, \ldots, N$, we can approximate the previous estimator by:

$$\hat{\mathbb{E}}_p[\mathcal{F}(\mathbf{x}_{0:t})] = \frac{1}{N}\sum_{i=1}^{N} \mathcal{F}(\mathbf{x}_{0:t}^{(i)})\frac{p(\mathbf{x}_{0:t}^{(i)})|\mathbf{y}_{1:t})}{q(\mathbf{x}_{0:t}^{(i)}|\mathbf{y}_{1:t})} \quad [1.6]$$

The law of large numbers allows us to show that this estimator almost certainly converges toward $\mathbb{E}_p[\mathcal{F}(\mathbf{x}_{0:t})]$ when N tends to infinity. Thus, we define the importance weights by $w_t^{*(i)} = \frac{p(\mathbf{x}_{0:t}^{(i)}|\mathbf{y}_{1:t})}{q(\mathbf{x}_{0:t}^{(i)}|\mathbf{y}_{1:t})} = \frac{p(\mathbf{y}_{1:t}|\mathbf{x}_{0:t}^{(i)})p(\mathbf{x}_{0:t}^{(i)})}{p(\mathbf{y}_{1:t})q(\mathbf{x}_{0:t}^{(i)}|\mathbf{y}_{1:t})}$, whose expression requires the calculation of the integral $(p(\mathbf{y}_{1:t}) = \int_{\mathbf{x}_{0:t}} p(\mathbf{y}_{1:t}|\mathbf{x}_{0:t})p(\mathbf{x}_{0:t})d\mathbf{x}_{0:t})$, which is generally impossible. We can nevertheless show that the following equation is usable [DOU 01]:

$$\begin{aligned}\hat{\mathbb{E}}_p[\mathcal{F}(\mathbf{x}_{0:t})] &= \frac{1}{N}\sum_{i=1}^{N} \mathcal{F}(\mathbf{x}_{0:t}^{(i)})\frac{w_t^{(i)}}{\sum_{j=1}^{N} w_t^{(j)}} \quad \text{with} \\ w_t^{(i)} &\propto \frac{p(\mathbf{y}_{1:t}|\mathbf{x}_{0:t}^{(i)})p(\mathbf{x}_{0:t}^{(i)})}{q(\mathbf{x}_{0:t}^{(i)}|\mathbf{y}_{1:t})}\end{aligned} \quad [1.7]$$

This estimator almost certainly converges when N tends to infinity. Then, it is sufficient to make the importance sampling recursive, to obtain the particle filtering algorithm described below.

1.2.2.2. *Particle filter*

The idea is thus to represent and to approximate empirically the *a posteriori* density by a weighted sample of size N $\{\mathbf{x}_{0:t}^{(i)}, w_t^{(i)}\}$, $i = 1, \ldots, N$ such that:

$$p(\mathbf{x}_{0:t}|\mathbf{y}_{1:t}) \approx \frac{1}{N}\sum_{i=1}^{N} w_t^{(i)}\delta_{\mathbf{x}_{0:t}^{(i)}}(\mathbf{x}_{0:t}) \qquad [1.8]$$

where the individuals $\mathbf{x}_{0:t}^{(i)}$, also called particles, are the realizations of the random variable $\mathbf{x}_{0:t}$ (state of the object) in the state space (δ being the Dirac function). Every particle is therefore a possible solution of the state to approximate and its associated weight represents its quality according to the available observations. Hence, the sample $\mathcal{S}_t = \{\mathbf{x}_{0:t}^{(i)}, w_t^{(i)}\}_{i=1}^N$ at the instant t is calculated from the previous sample $\mathcal{S}_{t-1} = \{\mathbf{x}_{0:t-1}^{(i)}, w_{t-1}^{(i)}\}_{i=1}^N$, so as to obtain an approximation (via sampling) of the filtering density $p(\mathbf{x}_{0:t}|\mathbf{y}_{1:t})$ at the current instant. For this, three stages are necessary: i) a state exploration stage, during which we propagate the particles via the proposal function, ii) a stage for the evaluation (or the correction) of the particle quality, which aims to calculate their new weight and finally iii) an optional stage for particle selection (re-sampling). The generic particle filtering scheme (SIR filter – *sequential importance resampling*), between the instants $t-1$ and t, is summarized in the algorithm below.

1) Representation of the filtering density $p(\mathbf{x}_{0:t-1}|\mathbf{y}_{1:t-1})$ with a set of particles $\{\mathbf{x}_{0:t-1}^{(i)}, w_{t-1}^{(i)}\}$, $i = 1, \ldots, N$.

2) Propagation, or exploration of the state space, with an importance (or proposal) function:

$$\mathbf{x}_t^{(i)} \sim q(\mathbf{x}_t|\mathbf{x}_{0:t-1}^{(i)}, \mathbf{y}_{1:t}) \qquad [1.9]$$

3) Correction, or evaluation of the particle quality, with observations, by calculating the weights:

$$w_t^{(i)} = \frac{p(\mathbf{x}_{0:t}^{(i)}|\mathbf{y}_{1:t})}{q(\mathbf{x}_{0:t}^{(i)}|\mathbf{y}_{1:t})}$$

$$\propto \frac{p(\mathbf{y}_t|\mathbf{x}_t^{(i)})p(\mathbf{x}_t^{(i)}|\mathbf{x}_{t-1}^{(i)})p(\mathbf{x}_{0:t-1}^{(i)}|\mathbf{y}_{1:t-1})}{q(\mathbf{x}_{0:t}^{(i)}|\mathbf{y}_{1:t})} \qquad [1.10]$$

Assuming that $q(\mathbf{x}_t^{(i)}|\mathbf{y}_{1:t}) = q(\mathbf{x}_t^{(i)}|\mathbf{x}_{0:t-1}^{(i)}, \mathbf{y}_{1:t})q(\mathbf{x}_{0:t-1}^{(i)}|\mathbf{y}_{1:t-1})$, we have:

$$w_t^{(i)} \propto \frac{p(\mathbf{y}_t|\mathbf{x}_t^{(i)})p(\mathbf{x}_t^{(i)}|\mathbf{x}_{t-1}^{(i)})p(\mathbf{x}_{0:t-1}^{(i)}|\mathbf{y}_{1:t-1})}{q(\mathbf{x}_t^{(i)}|\mathbf{x}_{0:t-1}^{(i)}, \mathbf{y}_{1:t})q(\mathbf{x}_{0:t-1}^{(i)}|\mathbf{y}_{1:t-1})}$$

$$\propto w_{t-1}^{(i)} \frac{p(\mathbf{y}_t|\mathbf{x}_t^{(i)})p(\mathbf{x}_t^{(i)}|\mathbf{x}_{t-1}^{(i)})}{q(\mathbf{x}_t^{(i)}|\mathbf{x}_{0:t-1}^{(i)}, \mathbf{y}_{1:t})} \qquad [1.11]$$

The weights are then normalized $\tilde{w}_t^{(i)} = \frac{w_t^{(i)}}{\sum_{j=1}^N w_t^{(j)}}$.

4) Approximation of the filtering distribution expectancy, the *a posteriori* law at instant t:

$$\mathbb{E}(\mathcal{F}(\mathbf{x}_{0:t})) \approx \frac{1}{N}\sum_{i=1}^{N} \tilde{w}_t^{(i)} \mathcal{F}(\mathbf{x}_{0:t}^{(i)})$$

5) Resampling (if necessary).

The equations below allow us to approximate the trajectory of the objects, but they can also allow to approximate only their state at instant t, by simply integrating over $\mathbf{x}_{0:t-1}$. In practice, this amounts to replacing $\mathbf{x}_{0:t}$ and $\mathbf{x}_{0:t-1}$, respectively, by $\mathbf{x}_t$ and $\mathbf{x}_{t-1}$ in the algorithms. In the rest of this work, depending on the applications, either one or the other possibility will be studied.

Once the theoretical framework is defined, we will discuss the problem of visual tracking by particle filtering in the next section.

1.2.3. *Application to visual tracking*

The PF has been used in numerous disciplines, such as communication, networks, biology, economy, geoscience, social sciences, etc. In image processing, it has been used in many domains (medical imagery, video analysis, meteorological imagery, robotics, etc.), for various applications such as segmentations or tracking in video sequences, which is the primary subject of our research.

Visual tracking poses many problems, among which the changes in appearance or illumination, occlusion, the appearance and the disappearance of objects, environmental noise and erratic movements are just a few examples. Particle filtering allows us to represent the arbitrary densities, focusing on specific regions of the state space and managing multiple models. It is easy to implement, robust to noise and to occlusions, although this requires taking a certain amount of precautions, among which:

– the choice of the state model $\mathbf{x}_t$, defined by a set of information that characterizes the object to track;

– the choice of observations $\mathbf{y}_t$, which allow identifying the object to track;

– the definition of an importance (or proposal) function q to propagate particles in a way that will guide the search in the state space;

– the definition of a likelihood function $p(\mathbf{y}_t|\mathbf{x}_t)$, which will link the current state of the object to the observation;

– the choice of the resampling method in order to avoid the problem of degeneration, which we will explain further in this section.

We will later give several solutions suggested by the literature to each of these points.

1.2.3.1. *State model*

The choice of a model x_t for the state depends on the available knowledge and the characteristic differences of the object that we would like to track. In this part, we describe how to model the state x_t of an object.

The most common method to represent an object is to use its geometric characteristics, in particular its position in the image (this is the case of the illustration in Figure 1.3). The $2D$ form can be given by a set of arbitrary points [ARN 05a, ARN 07, VER 05b] or specific points, such as edges [DOR 10, DU 05], contour points [CAR 10, CHE 01, LAK 08, MOR 08, XIA 08] or reference points [TAM 06]. Classical forms are also used, such as rectangles [BRA 07a, HAN 05b, HU 08, LEI 06, LEI 08, PÉR 02, WAN 09] or ellipses [ANG 08, MAG 09, NUM 03a], as well as forms interpolated by splines [LAM 09, LI 04a, LI 03]. We can also use level-sets [AVE 09, RAT 07a] or active contours [RAT 05, RAT 07b, SHE 06]. Finally, more evolved models integrating the relations between sets of pixels [HOE 10, HOE 06] are sometimes used. Among $3D$ forms, we use simple shapes (parallelepipeds, spheres) [GOY 10, MIN 10, MUÑ 10, ROU 10], thin $3D$ mesh of the face [DAI 04, DOR 05], the human body [GAL 06] or the hand [BRA 07c, CHA 08], as well as the contours [PEC 06].

Recently, numerous studies were conducted on the tracking of articulated objects, in which an object was modeled by a set of $2D$ or $3D$ shapes linked between themselves by articulations [BER 06, BRU 07, QU 07, SIG 04, YU 09]. The appearance models are also used, which require learning color [MAR 11, WAN 07], thumbnails [BHA 09], illumination [BAR 09, SMA 07], the exposure [WAN 05] or multiple shapes [BRA 05, GIE 02]. We also find more exotic appearance models using blur [SMA 08] or laser [GID 08] information. Finally, the state can be described by movement information, given by the refined transformations [GEL 04, KWO 08, MEI 09], velocity and/or acceleration [BLA 99a, CUI 07, VER 05a, DAR 08b] (we sometimes talk about auto-regressive models) or the trajectory [BLA 98a].

Naturally, these models are often combined to improve the description of the object, which increases the size of the state space, often making calculations unacceptable. We then need to make a compromise between the quality of the description and the computation time. Figure 1.1 gives several examples of state models used in tracking by PF.

1.2.3.2. *Observation model*

Here, again, the choice of the observation model y_t depends on the available information. In visual tracking, this information is extracted from the images, which are generated by different types of sensors, the number of which can vary. Many approaches work directly on pixels, which are often filtered during a simple pre-processing stage [BHA 09, GEL 04, GON 07, KAZ 09, KHA 06, SCH 07] or simply on pixels of the area from the extracted foreground [CHE 03a]. The difference between these approaches depends on the form of acquisition, which can supply, for example, fluorescent [LAM 09], $2D$ [SHE 06, SMA 08] or $3D$ [CHE 08] microscopic, infrared [PÉT 09] or even ultrasound [SOF 10] imagery. Note that for the color, we primarily use RGB representations [CZY 07, HU 08, MAG 05a, MAG 07, MAR 11, NUM 03a] and HSV [LIU 09, MUN 08b, PÉR 02, PER 08, SNO 09] (the latter being generally more adapted to vision problems, as it is less sensitive to changes in illumination). Other types of sensors are sometimes used, providing information such as distance and depth maps [ARN 05a, BER 06, LAN 06, MUN 08b, ZHU 10], movement maps [SCH 06], laser data [CUI 07, GID 08, GOY 10], projective images [ERC 07], occupation [MUÑ 09] or sound [CHE 03a, PÉR 04] maps. Figure 1.2 gives several examples of these.

Figure 1.1. *Some examples of state models used to represent the object to track. Form left to right, top to bottom, a model integrating illumination [BAR 09], an articulated model [SIG 10a], a trajectory [BLA 98a], a 3D facial mesh [DOR 05], level sets [AVE 09], a sphere [ROU 10], a set of points-of-interest [ARN 07], areas and their relations [HOE 10], a rectangle [BRA 07a], edges [DOR 10], an ellipsis [MAG 09] and appearance models [MAR 11]. For a color version of the figure, see www.iste.co.uk/dubuisson/tracking.zip*

1.2.3.3. *Importance function*

The importance function, or the proposal function, $q(\mathbf{x}_t|\mathbf{x}_{0:t-1}^{(i)}, \mathbf{y}_{1:t})$ makes it possible to guide particles, between two instances, in the state space in order to have *a priori* approximation of the density of the tracked object. Its choice is essential, since if the particles are propagated in inappropriate areas, the tracking will fail.

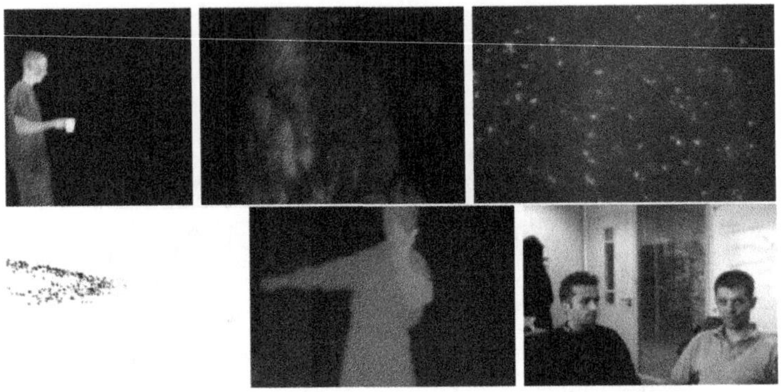

Figure 1.2. *Several examples of images modalities that are used as observations for tracking by PF. From left to right: infrared [PÉT 09], ultrasound [SOF 10], 2D microscopy [SMA 08], occupation map [MUÑ 09], depth map [ZHU 10] and Charge-Coupled Device (CCD) [PÉR 04]. For a color version of the figure, see www.iste.co.uk/dubuisson/tracking.zip*

The most common choice for the importance function is the transition function $p(\mathbf{x}_t|\mathbf{x}_{t-1})$. In this case, the importance weight is proportional to the likelihood function, as equation [1.11] becomes $w_t^{(i)} \propto w_{t-1}^{(i)} p(\mathbf{y}_t|\mathbf{x}_t^{(i)})$. This filter is more commonly called *Bootstrap* or the CONDENSATION algorithm [GOR 93, ISA 98a]. Here, we do not use the *a priori* information on the changes between the two instants, which are often modeled by two random Gaussian steps around the previous state approximation (see example illustrated in Figures 1.3(d) and 1.3(f)) or auto-regressive models from the first to the second order, integrating information on kinematics (velocity, acceleration). We can also retrieve the transition from the past [ISA 98a]. The problem with sampling the transition function is that if the transition model is inappropriate, most particles generated via this model will be "lost", as they will not be corrected properly.

Figure 1.3(f) shows the areas of particle diffusion (in blue) and of observation characterization (in red) (see color version of the figure): if these areas do not overlap, then the tracked object will be lost. The necessity to use an optimal importance function for visual tracking becomes obvious once the object can perform sudden movements that we can neither anticipate not model.

Hence, numerous works attempted to best approximate the optimal importance function, which generates particles randomly, and which thus plays an essential role in particle filtering [DOU 01, PIT 99, MER 00]. It has been shown that the optimal importance function (in the sense of minimizing the variance of the sample) needs to integrate the last observation and that it is then written $q(\mathbf{x}_t|\mathbf{y}_t, \mathbf{x}_{t-1})$ [DOU 00b]. Unfortunately, in most computer visualization problems, this expression is unknown and we have therefore searched for other solutions using the current observation. For example, a mix between the classical transition density and the detection function defined by a learning algorithm is used in [LU 09]. Other proposal functions are deducted from learning, for example movement over time [SHO 11]. A simulation taking advantage of an approximation of a displacement field by optical flow of the scene has also been suggested in [ARN 07], and in the partially linear context, allows extracting the optimal proposal function.

Other works suggested migrating the particles toward regions with high likelihood. This is the case of auxiliary particle filtering [PIT 99], which pre-selects the particles to propagate according to their link with the most recent observation. The proposal function of each of the particles can also be defined by an extended [TAN 96] or unscented [JUL 97] Kalman filter, in order to approximate the optimal proposal function by a Gaussian probability density. In the case of likelihood sampling [TOR 04], the particles are sampled directly from the likelihood density.

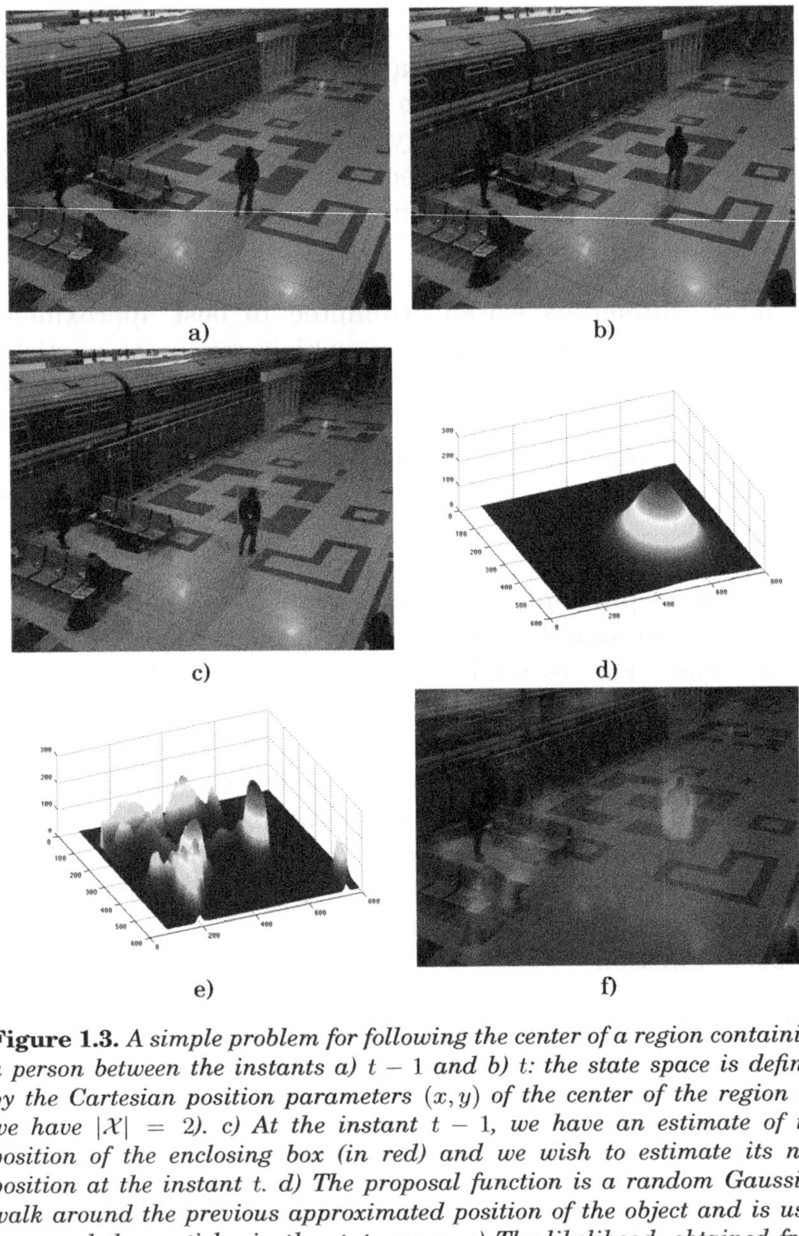

Figure 1.3. *A simple problem for following the center of a region containing a person between the instants a) $t-1$ and b) t: the state space is defined by the Cartesian position parameters (x, y) of the center of the region (so we have $|\mathcal{X}| = 2$). c) At the instant $t-1$, we have an estimate of the position of the enclosing box (in red) and we wish to estimate its new position at the instant t. d) The proposal function is a random Gaussian walk around the previous approximated position of the object and is used to spread the particles in the state space. e) The likelihood, obtained from the Bhattacharyya distance [BHA 43] between the color distribution in the enclosing box approximated at $t-1$ (image (c)) and in the target histograms, will be able to affect the weight of the particles. f) The influence areas of these two densities in the image space with the proposal function in blue and the likelihood function in red: the higher the tint of the pixel, the more value is attributed to the density. For a color version of the figure, see www.iste.co.uk/dubuisson/tracking.zip*

In fine, there are many proposal functions whose definition is based on the characteristics of the image and that are often related to the problem or the context. Among these *ad hoc* solutions, we can mention the works suggested in [ISA 98b], where an auxiliary contour is used to generate samples, or those described in [PÉR 04], which detect salient points according to precise characteristics (color, movement) that are centered on normal distributions, thus modeling the proposal function by a Gaussian mixture. Finally, in [SUL 01], the particles are guided with the density supplied by the appearance model.

1.2.3.4. *Likelihood function*

The likelihood function gives us a reason to believe in the validity of the observation $\mathbf{y}_t$, given the state $\mathbf{x}_t$ of the object. The way that we perceive this belief leads to first represent synthetically the available information based on the current observation $\mathbf{y}_t$, and then calculate the difference by comparison to the synthetic representation of an ideal situation. For example, this difference $\mathcal{E}$ can measure the similarity or the distance between the model of the previously estimated state and the model of the target corresponding to a hypothesis (or particle). Particle filtering determines the probabilistic framework in which this belief is modeled by a probability density $p(\mathbf{y}_t|\mathbf{x}_t)$, and therefore a common definition of the likelihood function is given by $p(\mathbf{y}_t|\mathbf{x}_t) \propto e^{-\lambda\mathcal{E}^2}$, where λ is the deviation parameter that makes the likelihood more or less pronounced.

There are several detailed studies on likelihood, generally made in a specific context. For example, a study on the choice of measure for similarity to compare color models using histograms is provided in [DUN 10], while the influence of the deviation parameter λ is studied in [LIC 04]. The choice of the likelihood function is directly related to the information used to characterize the tracked object. The literature is abundant on this subject. We often use color models described by histograms, as laid out in the early articles [PÉR 02], as well as many others [JAW 06, NUM 03b, SAT 04, WU 08b], where the Bhattacharyya distance allows measuring the similarity of two histograms (see the illustrations in Figures 1.3(e) and 1.3(f)). Other measurements have also been used, such as the Jensen–Shannon divergence [PER 08], the diffusion distance [LU 09] and the Earth Mover's Distance (EMD) distance [KAR 11]. Likelihood is also based on other types of information, such as the contours [MAC 99a], the foreground [LEI 08], statistics on pixel [LAM 09, MAG 05b, PAL 08], the shape [CAR 10, RAT 07b], the texture [LAO 09], the appearance (combination of shape and texture) [HOE 10, LEI 06, ZHO 04] as well as characteristic features [ARN 05b, DU 05].

Naturally, with the amount of information, the current trend is to combine likelihoods, often by multiplying them, under the independence hypothesis, in order to be able to handle multi-object, multi-characteristic, multi-modal or multi-view tracking problems. Examples of these will be given in Chapter 2.

1.2.3.5. *Resampling methods*

The particle filter and its variants all use a resampling stage in order to avoid the issue of degeneration in particles, that is the cases where the weight of every particle except one is close to zero. In practice, the variance of the importance weight $w_t^{(i)}$ increases over time, which has dramatic consequences on the tracking. There are several resampling methods and their goal is always the same: duplicate particles with high weight and, implicitly, eliminate those with low weight. A theoretical comparison of their advantages and downsides is given in [DOU 05]. Here we outline five methods most used in tracking.

– Multinomial resampling [GOR 93] (the most used) consists of selecting N numbers k_i, $i = 1, \ldots, N$, according to a uniform distribution $\mathcal{U}(0,1)$. The sample $\mathcal{S}_t = \{\mathbf{x}_t^{(i)}, w_t^{(i)}\}$

is replaced by a new sample $\mathcal{S}'_t = \{\mathbf{x}_t^{(D(k_i))}, \frac{1}{N}\}$ where $D(k_i)$ is a unique integer j such that $\sum_{h=1}^{j-1} w_t^{(h)} < k_i \leq \sum_{h=1}^{j} w_t^{(h)}$. If $(n_1, \ldots, n_N)$ indicates the number of times when particles of $\mathcal{S}_t$ are duplicated, then $(n_1, \ldots, n_N)$ is distributed according to the multinomial law $\mathcal{M}(N; w_t^{(1)}, \ldots, w_t^{(N)})$ (the descendants of the particles conjointly follow a multinomial law). In other words, by sampling with replacement N times the probability $\mathcal{M}\left(1; w_t^{(1)}, \ldots, w_t^{(N)}\right)$, we obtain N new particles i.i.d. according to $p(\mathbf{x}_t|\mathbf{y}_{1:t})$, with a weight of $1/N$.

– Stratified resampling differs from multinomial resampling by randomly selecting k_i according to the uniform distribution $\mathcal{U}(\frac{i-1}{N}, \frac{i}{N})$.

– Systematic resampling [KIT 96] randomly selects a number k according to $\mathcal{U}(0, \frac{1}{N})$ and then defines k_i such that $k_i = \frac{i-1}{N} + k$.

– Residual resampling [LIU 98] is very efficient for reducing the variance of the set of particles inducted by the sampling state. First, for each $i \in \{1, \ldots, N\}$, $n'_i = \lfloor N w_t^{(i)} \rfloor$ duplicate particles $\mathbf{x}_t^{(i)}$ of $\mathcal{S}_t$ are inserted into $\mathcal{S}'_t$. The remaining $N - \sum_{i=1}^{N} n'_i$ particles are selected randomly according to the distribution $\mathcal{M}(N - \sum_{i=1}^{N} n'_i; N w_t^{(1)} - n'_1, \ldots, N w_t^{(N)} - n'_N)$, for example through multinomial resampling. The weights of the $\mathcal{S}'_t$ particles are all equal to $1/N$.

– Weighted sampling is defined as follows. Let $g : \mathcal{X} \mapsto \mathbb{R}$ be a strictly positive continuous function and ρ_t be the multinomial distribution defined by $\rho_t^{(i)} = g(\mathbf{x}_t^{(i)}) / \sum_{j=1}^{N} g(\mathbf{x}_t^{(j)})$ for $i = 1, \ldots, N$. Take $k_1, \ldots, k_N$, independently according to the probability ρ_t. We construct the set of particles $\mathcal{S}'_t = \{\mathbf{x}_t^{(k_i)}, w_t^{(k_i)} / \rho_t^{(k_i)}\}_{i=1}^{N}$. It has been shown in [MAC 00a] that $\mathcal{S}'_t$ represents the same probability distribution as $\mathcal{S}_t$, focused on the modes of g (the usual choice for g is therefore the likelihood function). Note that, in contrast to other methods, weighted resampling does not affect particles with a weight of $1/N$.

A problem may occur if the sample is resampled too often: the representation decays (*sample impoverishment*), as the highest importance weights are duplicated too many times and the sample is reduced to a single particle. Moreover, every resampling stage diminishes the statistical independence of the particles, which is a strong assumption necessary for the convergence of the filter. Therefore, the decision to resample must be made at an opportune moment, in order to avoid increasing the variance of the sample, as well as to maintain a reasonable number of "good" particles (i.e. with relatively high weights) over time. One solution is to resample when $N_t^{\text{eff}} = \left(\sum_{i=1}^{N} (w_t^{(i)})^2\right)^{-1}$ reaches a threshold value, that is when the amount of "good" particles becomes too small (in practice, often fixed at 75% of N).

1.3. Limits and challenges

There are many versions of the PF [CHE 03b, DOU 01, MAS 01] and its primary appeal is its capacity to process and represent arbitrary densities, maintain multiple hypothesis, take into account non-Gaussian noise and to focus on areas of state spaces. Furthermore, it is relatively simple to implement and extend, robust to "noisy" backgrounds and to occlusions, rendering it rather suitable for problems in vision. In spite of this, the PF suffers from a certain number of downfalls. One of the previously discussed issues is that of particle degeneration, i.e. the loss of particle diversity in a sample, which can only be solved through regular resampling, by using an optimal proposal function or, failing that, a function approaching the optimum, whose choice is therefore critical. The choice of the resampling frequency must also be made with caution. Finally, the optimal number N of particles is impossible to define, as it is mostly dependent of the targeted application.

Nevertheless, the major problem, which remains a large constraints in this methodology, is quite certainly the calculation complexity. Indeed, the amount of particles necessary for good tracking increases exponentially with the size of the spaces in which the state and observation models are defined [MAC 00b]. The first applications of particle filtering in the domain of our interest concerned tracking the position of objects corresponding to the center of an enclosing box [NUM 02, PÉR 02, RUI 01] or contours [MAC 99a, TOR 01], by considering sometimes complex schemes integrating several objects, occlusions or changes of appearance, as shown in Figure 1.4.

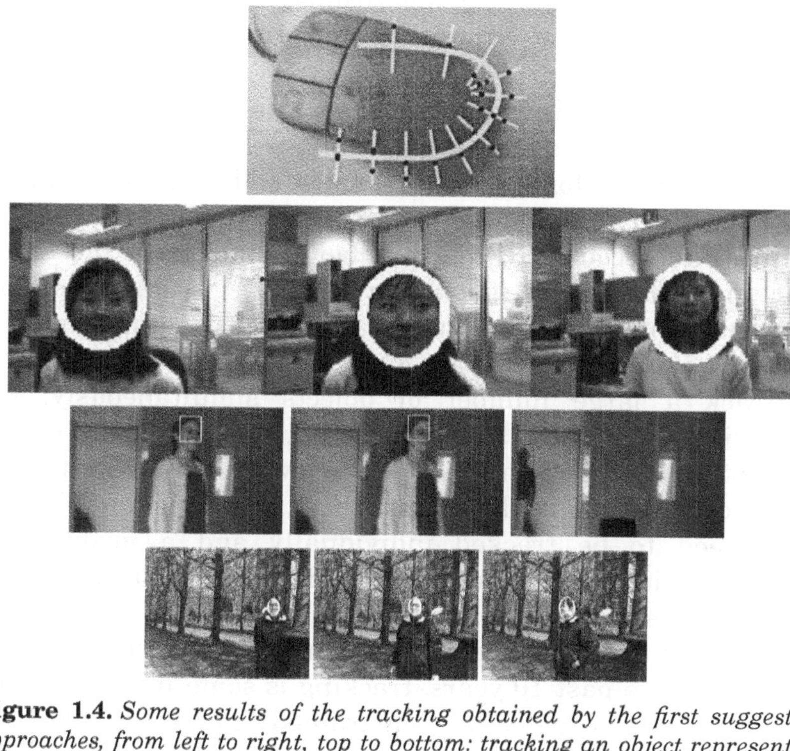

Figure 1.4. *Some results of the tracking obtained by the first suggested approaches, from left to right, top to bottom: tracking an object represented by its deformable contour [MAC 99a], tracking a face with changes in scale [RUI 01], tracking two faces with occlusion management [PÉR 02] and tracking a face with appearance changes [NUM 02]. For a color version of the figure, see www.iste.co.uk/dubuisson/tracking.zip*

The current tendency is to use as much information as possible for tracking, either by mixing different observation sources (several views, several forms of acquisition) or by refining the models that we use for the objects tracked, which requires increasing the number of parameters of the state vector and therefore the size of the space in which the latter is defined. To give an example, tracking human beings requires more and more precision and the models tend toward complex articulated objects similar to those in Figure 1.5(a), which entail a very large state space. The new representations of human movement [GUE 12], as well as the basis for testing tracking algorithms, such as HumanEva [SIG 10a], direct research toward characterization of posture, behavior recognition, gait analysis or even the detection of events such as falling over, which require more and more precise models. We also seek to have a fine analysis of movement and deformation, as illustrated in Figure 1.5(b) for the case of a deformable surface or interacting hands. Finally, taking into account a large number of objects in the same tracking scheme, in particular by integrating relations between them to improve their tracking, is also important. In Figure 1.5(c), we can see two applications for which we need to track a large number of objects simultaneously: for analyzing automatically the game tactics in a sport or for characterizing the behavior of a crowd. Currently, the PF does not allow doing this: when there are too many objects, they need to be tracked individually and a module for measuring interactions must be added as a post-process to make analysis. If a lot of progress aimed toward integrating increasingly complex models (see examples in Figure 1.1) was made over the past 10 years, tracking is sometimes slow and the solutions are often constrained to optimize their implementation (for example simplifyed hypotheses are used to bypass the complexity of certain calculation or models that can only be used for a certain type of application).

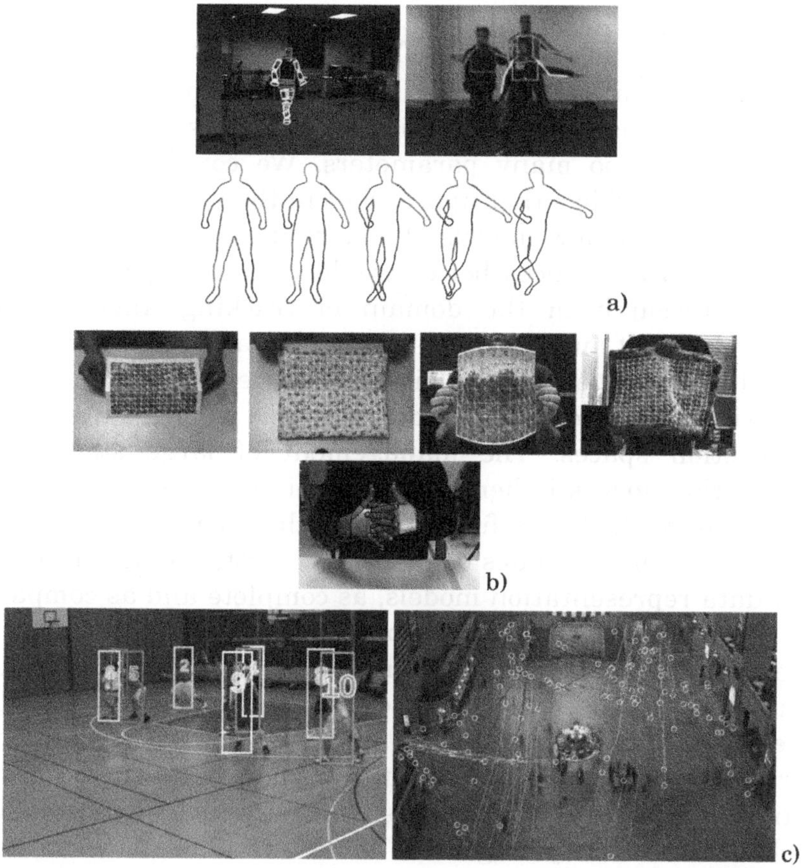

Figure 1.5. *Examples of complex tracking cases that still do not have an efficient solution in particle filtering. a) Complex modeling of a human body, from left to right by articulated object with a fine representation [SIG 10b], by multiple skeletons [BER 11] or by free and deformable shapes [FRE 10]. b) Tracking a deformable 3D surface [WAN 11a] and modeling the interaction between two hands [OIK 12]. c) Tracking multiple objects in a dense environment, from left to right are cases studied in [HUO 12] and [ZHO 12]. For a color version of the figure, see www.iste.co.uk / dubuisson / tracking.zip*

1.4. Scientific position

The application examples mentioned previously are currently infeasible with particle filtering, as the modeling still requires too many parameters. We could stop at the results provided by other approaches in the domain; however, we believe that new models of PFs that manage large state and observation spaces better would offer new perspectives on the research in the domain of tracking. Indeed, the management of multiple hypothesis naturally integrated into this filter could be fully exploited, if we were not constrained to using representations defined in small state and observation spaces. The management of large state and observation spaces is hence, in our opinion, a major current challenge to tackling future research problems. For this purpose, several avenues can be used. First, the definition of new data representation models, as complete and as compact as possible, is essential. In other words, integrating the wealth of the representation resulting from multiple characteristics used for representing the object in particle filtering is the first avenue to follow. Second, as we stated, particles are spread in the state space with the aim to optimally sample the filter density. Another way to solve the problem of the very large size of the state and observation space is therefore to provide models that more cleverly cover these spaces. For this purpose, we can either focus the search in the state space, in order to only cover interesting areas, or decompose it in sub-spaces, where sub-calculations can be made. We provide solutions using these different avenues in Chapters 2, 3 and 4 of this book.

1.5. Managing large sizes in particle filtering

The algorithms that tackled the problem of large sizes of state spaces can be roughly divided into three primary classes: those that reduce the size of the space, often by adding constraints to the model, those that use a local search and finally those that decompose the space into sub-spaces of a lesser size.

The first class of approaches adds constraints to the mathematical model in order to reduce the size of the state space to explore. In many articles, this is accomplished by introducing the physical constraints of the articulate object into the tracking model [BRU 09, OIK 11]. In particular, this type of approach is very popular in human tracking. Hence, in [VON 08], the constraints are introduced during a simulation stage, while in [BRU 07, HAU 10b], they are included directly in the proposal function constructed specifically to follow a person. Other approaches adding constraints add *a priori* information to the object [HAU 10a, COV 00, HAU 11], exploit the knowledge of its behavior [DAR 08a] or take into account its interaction with the environment [KJE 10].

The second class of approaches, often known as optimization oriented approaches, is a set of algorithms that combine particle filtering with local search techniques [MIN 86]. Given the stochastic nature of the filter and the combinatorial size of the state spaces of the objects, the PF is never guaranteed to produce a set of particle positions sufficiently close to the modes of the density to approximate. Thus, combining the filter with local search techniques can improve significantly its capacity to focus on these modes. For this reason, optimization approaches are very popular within the community working on object tracking. Among these techniques, we can mention the gradient descent methods that were specifically studied in this context. For example, stochastic gradient descent was successfully combined with particle filtering [HOF 11] and new stochastic meta-descent approaches were suggested in a constrained space [BRA 07b], giving the efficient *smart particle filter* [BRA 7d]. Particle swarm optimization techniques are also used conjointly with the filter [WAN 06, LI 11, JOH 10, KRZ 10]. Here, the idea is to apply evolutionary algorithms inspired by social animal behavior (birds, fish, bees, ants, etc.) to evolve the particles in accordance with their own experience, as well as that of their neighbors (for a complete review of these techniques, we recommend the lecture of [DAS 11]). Similarly, the introduction of population-oriented meta-heuristics and genetic algorithms was used in the context of particle filtering [SÁN 05a]. Simulated annealing was also introduced into particle filtering, giving the renowned *annealed PF* (APF) (or particle filtering with simulated annealing) [DEU 05]. APF adds iteration of pseudo-simulated annealing to the resampling, in order to spread the particles in the state space and hence position them closer to the modes of the density to approximate. Naturally, there are other optimization-oriented methods (such as the scatter search [PAN 08a], etc.) that we cannot list here, as they are not the main subject of this work. Nevertheless, all of these approaches share the commonality of being a compromise between the quality of the approximation of the estimated density and the velocity of convergence. Unfortunately, given their local nature, although many of these approaches may converge rapidly to a local minimum near their starting point, they require a lot more time to converge to the overall minimum, which, in addition, is not systematically guaranteed. Hierarchic approaches tried to resolve these problems by adopting a strategy of progressively refining the search space, starting from a general description of the state space to end on a finer one, giving a complete description. The *progressive particle filter* [CHA 10] is an example of this. Finally, it is important to mention that all of these methods assume that every required observation is available at every instant, which is unfortunately not necessarily the case in practice.

The third class of approaches exploits the natural decompositions of state and observation spaces into a set of sub-spaces of a more reasonable size, where particle filtering can be applied. Partitioned sampling (PS) [MAC 99b] is probably the best known here. It uses the fact that in many problems, both the dynamics and the likelihood can be decomposed. The key idea is to substitute the application of the filter to the whole space with a sequence of applications to the sub-spaces, therefore significantly accelerating the process. Despite recent improvements [SMI 04, DUF 09], PS suffers from too high a number of necessary resamplings, which increases the noise, leading to a decrease in tracking quality over time. An equivalent type of decomposition is used in [KAN 95], in the context of dynamic Bayesian networks (DBNs). Here, the proposal density of the predictive stage is decomposed as a product of conditional distributions in each node of the DBN at the current instant. The predictive stage is then executed iteratively on each node of the network, according to the topological order by using the proposal distribution of the node based on its parent in the network. In [ROS 08], the sampling idea suggested in [KAN 95] is combined with the resampling scheme from [MAC 99b] in order to create a particle filtering algorithm well adapted to DBN. This algorithm can be seen as a generalization of PS. By following the topological order of the network for sampling and resampling the particles every time a node is explored, the particles with low likelihood in a sub-space are excluded, while those with a high likelihood are multiplied. The effect is similar to that of the weighted resampling in PS. Another approach coming from the Bayesian community is the *non-parametric belief propagation* algorithm [SUD 10, ISA 03]. It combines particle filtering with the *loopy belief propagation* algorithm [PEA 88, YED 05], in order to accelerate calculations (at the price of worse approximations). It was successfully applied to the large-size problem [SUD 04, SIG 03, BER 09, IHL 04, LAN 06]. Another popular approach is the *Rao-Blackwellized PF* (RBPF) *for DBN* [DOU 00a]. By using the natural decomposition of joint probability, RBPF decomposes the state space into two parts, following these conditions: the distribution of the second part, based on the first, can be estimated with a Kalman filter. The distribution of the first part is estimated by particle filtering. As the size of the first part is smaller than the whole space, its sampling stage needs less particles, and the variance of the estimation can therefore decrease. Although RBPF is very efficient at reducing the size of the problem, it cannot be applied to any Bayesian network, as the state space cannot always be decomposed in the manner that it assumes (i.e. assuming a part of the space to be linear). The work suggested in [BES 09] is a parallelized PF dedicated to Bayesian networks, which uses the same probabilistic decomposition of the joint distribution in order to reduce the number of particles required for tracking. The state space is divided into a set of independent sub-spaces. The particles are hence generated independently in the sub-spaces according to different proposal densities. This approach offers real flexibility in terms of proposal density choices. Nevertheless, their definition requires the underlying Bayesian network to have certain structural properties, which limits the generalization of this algorithm.

1.6. Conclusion

We presented in this chapter the elements fundamental to the introduction and the definition of sequential Monte-Carlo methods, as well as their use in the context of tracking in video sequences. In particular, we have shown to which point the community was active in tracking with PF, making it evolve very fast. Nevertheless, this also allowed us to highlight a certain number of challenges that motivated our research. Specifically, the management of large state and observation spaces, which we see as a major challenge to undertake in the years to come, caught our attention. Later in this work, we discuss several solutions that allow us to advance in this direction. These methodological fields affect, on the one hand, the modeling of the data to process, as well as its representation, and, on the other hand, the exploration of the state space by focusing or by decomposition.

2

Data Representation Models

2.1. Introduction

In many computer vision problems, we need to make a compromise between the quantity of processed information and the quality of the results that we need to obtain. When we consider the problem of tracking, we encounter a multitude of data coming either from the representation chosen for the objects (color distribution, silhouettes, contours, etc.), which needs to be as precise as possible, or the quantity and quality of the available observations (multiple modalities, multiple views, etc.), which increase continuously. Hence, this requires modeling the state or the observation through a large number of parameters. Yet, every parameter added to the state vector of the object increases the dimension of the space in which this object is defined; more importantly, as we have mentioned in Chapter 1, this also exponentially increases the number of particles which are necessary for good tracking. This makes probabilistic computations, in particular marginalization (which requires integrations), impossible to compute. Likewise, every measure added to the observation makes the likelihood computations more complex. Works on data representation follow two primary directions. The first orientation, which is the topic of section 2.2, concerns the work on accelerating the computation of the likelihood function when the weights of the particles are computed from distances between histograms. The second orientation, described in section 2.3, is about the representation of more complex multifeature data and the modeling of classical computer vision problems (deformations and appearances/disappearances) and their natural integration into particle filtering to limit the exponential increase of the number of particles.

2.2. Computation of the likelihood function

In this section, we assume that the correspondence between a particle and the observation is measured by comparing distributions approximated by histograms. The colors or the gradient orientations are some of the information that can be represented in this way, and we will often use the Bhattacharrya distance as comparison measure $\mathcal{E}$. Hence, as we defined in section 1.2.3.4, the likelihood function is approximated for each particle $\mathbf{x}_t^{(i)}$, by $p(\mathbf{y}_t|\mathbf{x}_t^{(i)}) \propto e^{-\lambda\mathcal{E}^2}$.

DEFINITION 2.1 (Histogram and quantified histogram).– *The histogram H of an image $\mathcal{I}$ of size $W \times H$ is defined by $H(k) = \sum_{x=1}^{W}\sum_{y=1}^{H}\{\mathcal{I}(x,y) = k\}$, where $\mathcal{I}(x,y)$ is the value of a pixel (grayscale, gradient orientation, etc.). Quantifying the distribution of probability represented by H allows us to "summarize" it in B intervals, called* bins, *and the quantified histogram is defined by:*

$$H(b) = \sum_{x=1}^{W}\sum_{y=1}^{H}\left\{\mathcal{I}(x,y) \in \left[b\frac{K}{B}, (b+1)\frac{K}{B}\right[\right\} \quad b = 0, \ldots, B-1$$

where K is the dynamic-range of the values that a pixel can take.

The classical approach allowing us to compute the histogram of a region R of size $W \times H$ consists of traversing the set of its pixels, which leads to a time complexity of $O(W \times H)$.

In tracking by particle filter, every object, modeled by the state vector $\mathbf{x}_t$, can be identified with one or more histograms (for an articulated or a fragmented object). If we have N particles $\mathbf{x}_t^{(i)}$ possible state realizations, and the object is represented by a single histogram, we will have N histograms to compute, or $M \times N$ histograms if the object is represented by M histograms. The computations are very long (in particular, if N is large) and represent nearly 70% of the total computation time necessary for estimating density by particle filter: it is thus important to be able to accelerate them. For this purpose, we suggested four algorithms. The first two algorithms, detailed in section 2.2.1, exploit the spatial redundancy of the pixel values to accelerate the computations. The other two algorithms, explained in section 2.2.2, exploit the temporal redundancy to accelerate the computation of histograms and their distance.

In what follows, we consider the set of rectangular regions $\{R^1, \ldots, R^N\}$ (here, one per particle), without rotation in the plane, but whose size can vary and of which we would like to extract histograms $\{H_{R^1}, \ldots, H_{R^N}\}$, that we will then call *target histograms*. Every region R^i is identified by the quadruplet $\langle x^i, y^i, w^i, h^i \rangle$ where (x^i, y^i) are the coordinates of the lower right corner of the region and w^i and h^i represent, respectively, its width and height.

2.2.1. *Exploitation of the spatial redundancy*

The classical approach for computing histograms, which we described previously, aiming to compute the histograms by scanning the pixels of the regions, is acceptable if we only have a reduced number of histograms to compute. Unfortunately, certain applications, in particular tracking by particle filter, require the computation of a large number of histograms and, in this case, it is necessary to use a different approach, which remains reasonable in terms of computation time. To this end, we can observe that rectangular regions in which histograms are computed often overlap, leading to a partial redundancy (see example in Figure 2.1(a)). This redundancy has been exploited by a certain number of works to accelerate the computation of histograms. The simplest approaches [PER 07, TAN 79] were suggested for spatial filtering by moving window: the previously computed histogram is updated by adding and subtracting columns/rows following the movement of the window. A computation of the histogram using the image distribution property was also suggested in [SIZ 08]. This approach, which has proved itself faster than the previous approach, has the advantage of being easily extensible to the case of non-rectangular regions.

However, when a high number of histograms need to be computed, the *Integral Histogram* (IH) [POR 05] is today the most commonly used approach in real-time applications [ADA 06]. Before computing the target histograms, IH efficiently computes a set of *temporary histograms* (THs). More formally, we consider the smallest rectangular region without rotation containing the set of regions R^i, which we call the region of interest (ROI). IH computes, for every pixel (x, y) of the ROI, the $TH(x, y)$ of the region where (x, y) is the lower right corner. For this, the property of integral images is exploited for every couple of coordinates (x, y) of the ROI, which gives:

$$TH(x, y) = \mathcal{I}(x, y) + TH(x - 1, y) + TH(x, y - 1) \\ -TH(x - 1, y - 1). \quad [2.1]$$

Hence, the THs are computed from the upper left corner to the lower right corner, by scanning the ROI and using the already computed histograms, which makes the computation very efficient, as it only uses three arithmetic operations.

$$
\begin{aligned}
H_{R^i} = {} & TH(x^i, y^i) - TH(x^i - w^i, y^i) - TH(x^i, y^i - h^i) \\
& + TH(x^i - w^i, y^i - h^i).
\end{aligned} \tag{2.2}
$$

Here again, three arithmetic operations are necessary, which allow us to compute very quickly any target histogram from the THs.

The major disadvantage of IH is that the computation time of THs and the memory space required for their storage can become very large: if the size of the ROI is $W \times H$ and the histograms are quantified over B intervals, then the memory size necessary is $W \times H \times B$ octets. Moreover, the classical approach becomes more interesting to use in terms of computation time if we only need to compute a small number of histograms $\{H_{R^1}, \ldots, H_{R^N}\}$, as the speed of target histogram computation by IH does not compensate the time lost computing the set of THs.

In the following sections, we give two approaches that exploit spatial redundancies and show themselves to be more competitive in relation to the IH computation technique.

2.2.1.1. *Optimal order for histogram computation*

The idea suggested in [BER 12] is illustrated in Figure 2.1(b): by symbolizing the regions as a heap, we seek to determine the optimal order to use for computing their histograms, minimizing the costs by exploiting spatial redundancy. Hence, in this figure, we first compute the histogram of the red region (the "highest" one), then that of the green region from the one of the red region and finally, that of the blue region, based on the one of the green region.

An optimal order in which the histograms of the regions need to be computed is obtained by determining the *Minimum Spanning Tree* (MST) of the graph modeling the overlaps between the regions. Let V be the set of N points $2D$ $\mathrm{p}^i = (x^i - w^i/2, y^i - h^i/2)$, centers of the regions R^i, from which we are looking to extract histograms. Our approach decomposes into three main stages (see Figure 2.1(c) for an illustration, and the general framework in Algorithm 2.1):

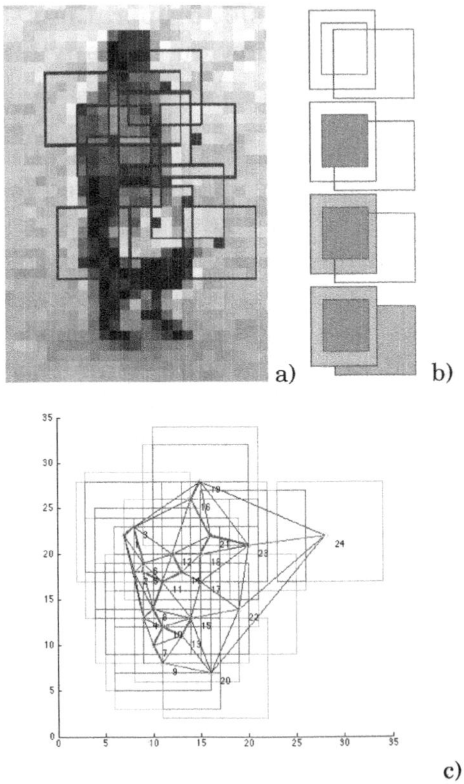

Figure 2.1. *Illustration of the approach suggested in [BER 12]. a) A classical case of overlapping regions in an application in tracking. b) Idea for $N = 3$ regions, from top to bottom: the three regions then the complete computation of the histogram of the red region, then that of the green region, by using the one of the red region, then that of the blue region, by using the one of the green region. c) Detail for $N = 30$ regions: a Delaunay triangulation (represented in blue) is first constructed from the set V of the centers of regions (green circles), to which is applied Kruskals algorithm in order to obtain a minimum spanning tree, in red (the thick red lines represent the edges for which the overlapping criterion c between two centers that they link is strictly positive). For a color version of the figure, see www.iste.co.uk/dubuisson/tracking.zip*

1) Construct the Delaunay triangulation $\mathcal{G} = (V, E)$ from the points p^i. The edges E are weighted by the distance (L_1 norm) between the centers that they link. We call w the set of the edge weights.

2) Compute the MST $\mathcal{T}$ of $(\mathcal{G}, \mathrm{w})$ by using Kruskal's algorithm [KRU 56].

3) Make a depth-first traversal of the $\mathcal{T}$ to obtain a path linking the centers of the regions in an optimal order for computing histograms. To this end, as the traversal indicator, we use a criterion representing the overlapping between two regions R^i and R^j,

$$c(R^i, R^j) = \max\left(0; 1 - \frac{|R^i|^- + |R^j|^+}{|R^j|}\right)$$

where $|R^i|^-$ (respectively, $|R^i|^+$) corresponds to the number of pixels to subtract from (respectively, to add to) the region R^i and $|R^i|$ is the total number of pixels in this region. If $c(R^i, R^j) = 0$, the histogram of R^j needs to be constructed entirely, otherwise, it can be updated from the histogram of R^i.

Algorithm 2.1: Algorithm for computing histograms in an optimal order [BER 12]

Input : set of regions $\{R^i\}_{i=1}^N$, set of centers $V = \{(x^i - w^i/2, y^i - h^i/2)\}_{i=1}^N$
Output : set of histograms $\{H_{R^i}\}_{i=1}^N$
$\mathcal{G} = (V, E) \leftarrow$ Delaunay triangulation of V
for each $e_k = (\mathbf{x}^p, \mathbf{x}^q) \in E$ **do**
  $w_k \leftarrow \mathrm{distance}_{L_1}(\mathbf{x}^p, \mathbf{x}^q)$
$W \leftarrow \{w_k : e_k \in E\}$
$\mathcal{T} \leftarrow$ `Kruskal`(V, E, W)
Choose a node $v \in \mathcal{T}$
$H_{R^v} \leftarrow$ histogram of R^v computed by traversing all the pixels of R^v
`depth-first_traversal`$(\mathcal{T}, v, \{R^i\}_{i=1}^N)$ (Algorithm 2.2)

Algorithm 2.2: Algorithm of the depth-first traversal

Input : tree $\mathcal{T}$, nodes v, set of regions $\{R^i\}_{i=1}^N$
Output : set of histograms $\{H_{R^i}\}_{i=1}^N$
Mark v as explored
for each *neighbor* v_s *of* v *in* $\mathcal{T}$ **do**
  if v_s *has not been explored yet* **then**
    if $c(R^v, R^{v_s}) \neq 0$ **then**
      $H_{R^{v_s}} \leftarrow$ histogram of R^{v_s} computed from H^v
    otherwise
      $H_{R^{v_s}} \leftarrow$ histogram of R^{v_s} computed by traversing all the pixels of R^{v_s}
    `depth-first_traversal`$(\mathcal{T}, v_s, \{R^i\}_{i=1}^N)$

Tests have shown two important properties of this approach. First of all, it only marginally depends on the quantification of histograms and the size of the ROI (see Figures 2.2(a) and 2.2(d)). Then, the computation times increase slowly with the size of the regions in which the histograms are computed, which makes this approach more efficient when the regions are large (as the ROI will also be larger, IH would loose a lot of time computing THs), even if for a very high number, this approacher becomes slower than IH (see Figure 2.2(b)). Finally, in contrast to IH, of which this is the main limitation, this approach remains efficient even when the number of histogram to compute is small (see Figure 2.2(c)). Still, this approach is less efficient when we need to compute a lot of highly quantified histograms in very small regions. Nevertheless, for visual tracking applications, a reasonable quantification is necessary (generally $B = 8$ or $B = 16$) and the areas of the image need to be large enough for the information that we extract to be sufficiently descriptive: we, therefore, consider this approach to be a good compromise to that of the IH.

2.2.1.2. *Optimization of the integral histogram*

IH computes one histogram per pixel of the ROI in its accumulator, despite the fact that not all of these histograms are necessary. A new approach, *Min-Space Integral Histogram* (MSIH) [DUB 12a], computes and stores only those that are necessary.

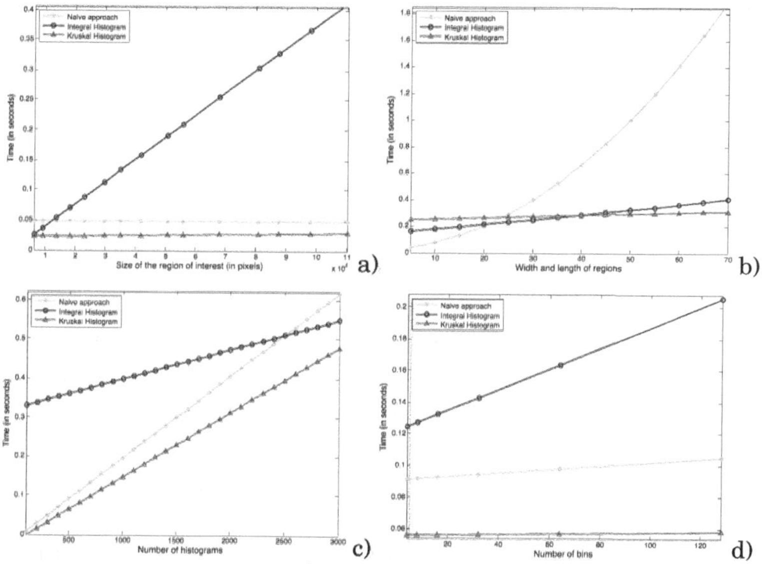

Figure 2.2. *Comparison between the computation of a set of histograms obtained by IH (in blue), the classical method (in green) and the suggested approach (in red), according to a) the size of the ROI (regions* 30×30*,* $N = 2000$*,* $B = 16$*), b) the size of the regions from which we extract the histograms (*$N = 2000$*,* $B = 16$*), c) the number* N *of histograms to compute (regions* 30×30*,* $B = 16$*, ROI* 270×250*) and d) the number* B *of intervals (regions* 30×30*,* $N = 500$*, ROI* 150×170*). For a color version of the figure, see www.iste.co.uk/dubuisson/tracking.zip*

The general idea of the MSIH is illustrated in Figure 2.3. Here, we wish to compute histograms of $N = 7$ regions, and we assume that the ROI is of the size 100×100. In this case, IH would need to store 10,000 THs, although only those of the regions whose lower right corners are blue (see color version of the figure) triangles or red circles (28 in total, see Figure 2.3(b)) are really necessary for computing the seven target histograms (see equation [2.2]). MSIH focuses only on those 28 THs. To this end, we determine a grid that contains these 28 points (gray lines in Figure 2.3(b)). Note that, in this example, this grid only counts 12 rows l and 14 columns c, for a total of 168 cells, instead of 100 rows x and 100 columns y, for a total 10,000 cells of the ROI. MSIH traverses this grid, much like IH traverses the ROI, from left to right and top to bottom, to compute incrementally the TH of the row l and the column c of the grid:

$$\begin{aligned} TH(l,c) = & \sum_{(x,y)\in G_{lc}} \mathcal{I}(x,y) + TH(l-1,c) + TH(l,c-1) \\ & -TH(l-1,c-1) \end{aligned} \quad [2.3]$$

where G_{lc} is the rectangular area of the image defined by the outer points $(G_x(l-1)+1, G_y(c-1)+1)$ and $(G_x(l), G_y(c))$, with $G_x(l)$ and $G_y(c)$ representing the coordinates (x,y) in the image, corresponding, respectively, to the row l and the column c of the grid. This scheme alone allows us to improve IH simultaneously on computation times (less arithmetic computations) and on memory consumption (less THs to store). Yet, this idea could be refined more, as it is not in fact necessary to compute all the $TH(l,c)$. We can effectively show that only those represented by the red (see color version of the figure) circles, blue triangles and green squares in Figure 2.3(c) are necessary to compute the 7 H_{R^i}. This reduces considerably the number of THs to compute: 91 instead of 168 cells of the grid in this example. Intuitively, the green squares are used to exploit the THs already computed in order to determine efficiently those of the red circles to their right. Therefore, they do not require additional storage, as we can use the storage memory of the histograms of these red circles to store them as they are needed. The green squares being used in the computations of the right borders of the regions R^i, they are only placed on the grid in places that are inside at least one region and there is a red circle to their right. Their identification, as well as the identification of the red circles and blue triangles, that is to say, the set of locations where histograms (especially temporary histograms) need to be computed, is given by Algorithm 2.3.

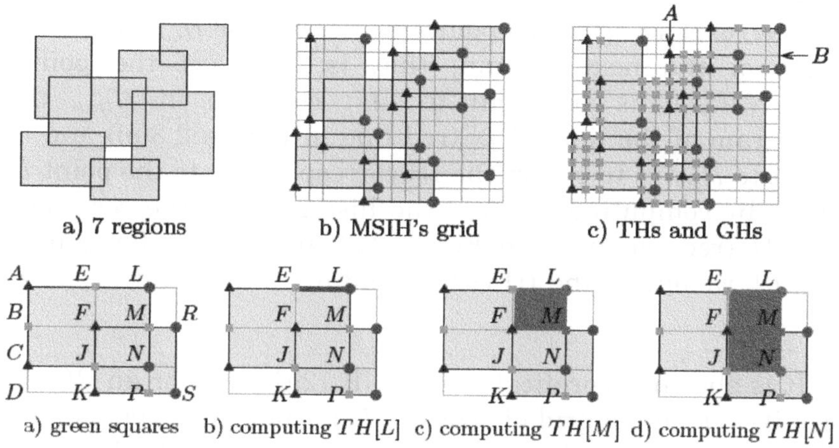

Figure 2.3. *Principle of the Min-Space Integral Histogram, MSIH [DUB 12a]. Top row, on the left are the seven target regions (gray rectangles) whose histograms we would like to obtain. In the middle, the lower right corners of the regions whose THs are the strict minimum (the blue triangles correspond to the left corners of the regions and the red circles to the right corners). On the right, the bottom corners of the regions of the computed THs (the square corners correspond to points where THs will be computed, but incur no extra memory cost), while IH computes them in every pixel of the ROI. Bottom row, principle of computing THs for a given column. For a color version of the figure, see www.iste.co.uk/dubuisson/tracking.zip*

Finally, the global process of TH computation is given by Algorithm 2.4. The lower row in Figure 2.3 illustrates the computation method. The idea is to traverse the grid column-by-column, form left to right and, within each column, from top to bottom, to update a vector H_{col} containing the set of all THs. As we can see in Algorithm 2.4, the size of this vector is at most the number of rows of the grid and, when no red circle remains in the row, the TH of this line is deleted. Hence, we can guarantee that the number of histograms created is never larger than $4N + 1$ (the 1 coming from an additional histogram H_{tmp} that allows us to compute equation [2.3] in two arithmetic operations, rather than the three used by IH). The bottom row of Figure 2.3 illustrates the procedure: let us assume that H_{col} contains the IHs corresponding to the points E, F, J, K. For H_{col} to contain those of the next column, that is, those of the points L, M, N, P, it is sufficient to first compute in H_{tmp} the histogram of the line $]E, L[$ (red line in Figure 2.3(e)) and to add it to $H_{col}[E]$ to obtain the IH corresponding to the point L. Then, by computing in H_{tmp} the histogram of the rectangle $ELFM$ (red rectangle in Figure 2.3(f)) and by adding it to $H_{col}[F]$, we obtain the IH corresponding to the point M, as so on.

It is shown theoretically in [DUB 12a] that, for an ROI of size $W \times H$ and a grid of size $w \times h$, at most $\min(wh; 4N) + 1$ histograms need to be stored, against WH for IH. Moreover, the complexity of this approach for the computation time of THs quantified over B intervals is in $O(W \log H + whB + WH)$, against $O(WHB)$ for IH.

This approach reduces significantly both the computation time of the histograms and the memory space necessary, in contrast to either IH or the classical histogram computation method. Tests showed that this approach allows us to compute up to 1,500 THs in less than IH and thus reduces up to 75 times the computation times. Furthermore, this approach remains more competitive than IH in the case where the number N of histograms to compute is small. This is of course due to the reduction of the number of computed THs. Figure 2.4 gives several examples of the tests conducted on the computation times by varying the quantification of the histograms, the number of regions or their overlap rate.

Algorithm 2.3: MSIH [DUB 12a]: identifying the lower right corners of the regions whose temporary histograms are computed

Input : regions $\{R^1, \ldots, R^N\}$, grid G of size $w \times h$
Output : set of points where temporary histograms should be computed

Initialize G: all cells are uncolored
// create 2 vectors V_T, V_B counting, for every line,
// the number of upper and lower corners of already processed regions R^i
$V_T, V_B \leftarrow$ vectors of size $h+1$ filled with 0
// fill the grid from right to left
for $l = w$ **at** 1 **do**
  // if there is a column to the right, create a green square:
  if $l \neq w$ **then**
    no_of_current_regions $\leftarrow$ 0
    for $c = 1$ **at** h **do**
      no_of_current_regions $\leftarrow$ no_of_current_regions $+ V_T[c] - V_B[c]$
      if no_of_current_regions > 0 **and** $G[l+1, c] \neq$ uncolored **then**
        $G[l, c] \leftarrow$ green square
  // fill G with the regions whose the right corners are in the column l
  $\mathcal{R} \leftarrow$ {regions R^k whose the right sides are in the column l of G}
  for each $R^k = \langle x^k, y^k, w^k, h^k \rangle \in \mathcal{R}$ **do**
    convert the image coordinates $(y^k, y^k - h^k)$ into grid coordinates (b^k, t^k)
    $G[l, b^k] \leftarrow$ red circle; $G[l, t^k] \leftarrow$ red circle
    $V_T[t^k] \leftarrow V_T[t^k] + 1$; $V_B[b^k + 1] \leftarrow V_B[b^k + 1] + 1$
  $\mathcal{L} \leftarrow$ {Regions R^k whose left sides are in the column l of G}
  for each $R^k = \langle x^k, y^k, w^k, h^k \rangle \in \mathcal{L}$ **do**
    convert the image coordinates $(y^k, y^k - h^k)$ into grid coordinates (b^k, t^k)
    $G[l, b^k] \leftarrow$ blue triangle; $G[l, t^k] \leftarrow$ blue triangle
    $V_T[t^k] \leftarrow V_T[t^k] - 1$; $V_B[b^k + 1] \leftarrow V_B[b^k + 1] - 1$

Algorithm 2.4: computation of temporary histograms

```
Input  : image 𝓘, grid G of size w × h
Output : set TH of temporary histograms computed at the blue and
         red points of G
H_col ← vector of size h of empty histograms
H_tmp ← empty histogram    TH ← ∅
for i = 1 at w do
    H_tmp ← ∅
    for j = 1 at h do
        if G[i, j] is a red circle, a blue triangle or a green square then
            Let R be the rectangle formed by the last rows and columns
            where temporary histograms were computed on the grid until
            G[i, j] (i.e., NJFM for H_col[3] (i.e. at point N) in
            figure 2.3.(g))
            add the pixels of the rectangle R to H_tmp
            add H_tmp to H_col[j]
            if G[i, j] is a blue triangle or a red circle then
                save H_col[j] in TH[i, j]
                if there is not red circle G[k, j], for k > i then
                    erase H_col[j] from memory
```

2.2.2. *Exploitation of the temporal redundancy*

The methods presented above aim to accelerate the computation of histograms by limiting as much as possible the repetition of computations and by exploiting spatial redundancies (see [SIZ 08] for a comparative study on this type of approach). However, other types of redundancies can be used, like the temporary redundancy, which has been exploited very little with the aim to accelerate the computation of histograms or likelihoods. This temporal redundancy occurs when the difference between the pixels of two consecutive images is minor (by "minor" we indicate the fact that the value of a pixel between two images is classed in the same histogram bin (or interval)). Here, we assume having computed the histograms or distances between histograms at instant $t - \delta$ and wish to use that to compute the instant t.

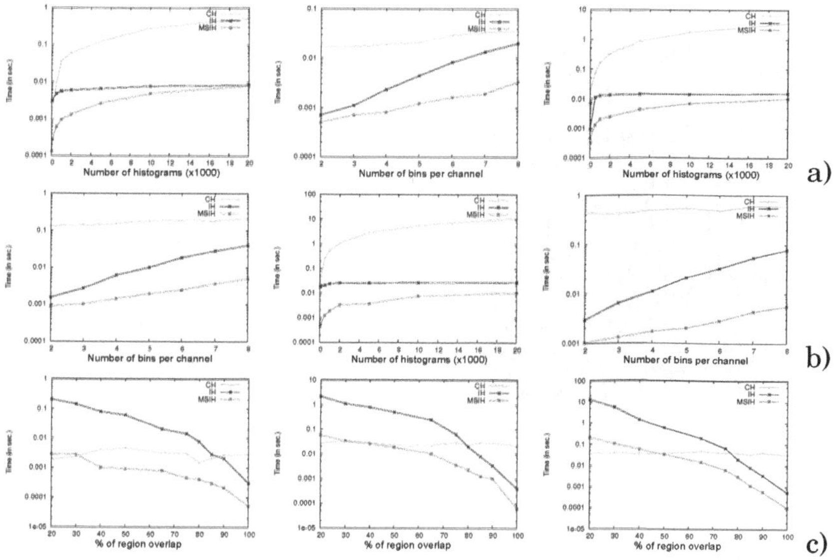

Figure 2.4. *Comparison of the computation times (in seconds, logarithmic scale) obtained by the classical method (CH, in green), the IH (IH, in blue) and our approach (MSIH, in pink) according to (a) the number N (multiple of 1,000) of regions (sizes from left to right $\{10 \times 10, 50 \times 50, 100 \times 100\}$, quantified RGB histograms $(6 \times 6 \times 4)$), (b) is the number of intervals by color channel (RGB), for $N = 1000$ regions (sizes from left to right $\{10 \times 10, 50 \times 50, 100 \times 100\}$) and (c) the percentage of overlapping between the regions of size 30×30 (from left to right $N = \{100, 500, 1000\}$, quantified RGB histograms $(6 \times 6 \times 4)$). For a color version of the figure, see www.iste.co.uk/dubuisson/tracking.zip*

2.2.2.1. *Temporal histogram*

The idea is to update between two instances the histogram $H_{R^i t}$ of a region R^i by only taking into account the significant differences that occurred in this region between the instants $t - \delta$ and t. From a simple image of differences between these two instants, we construct a tree of depth 3, which stores the information on temporal differences. Hence, for every pixel (x, y) changing between $t - \delta$ and t, we store the column x at depth 1, its row y at depth 2 and, at the level of the leaves, the previous and the new pixel interval, respectively, $b_{t-\delta}$ and b_t. Figure 2.5 gives an example illustration of a simple case.

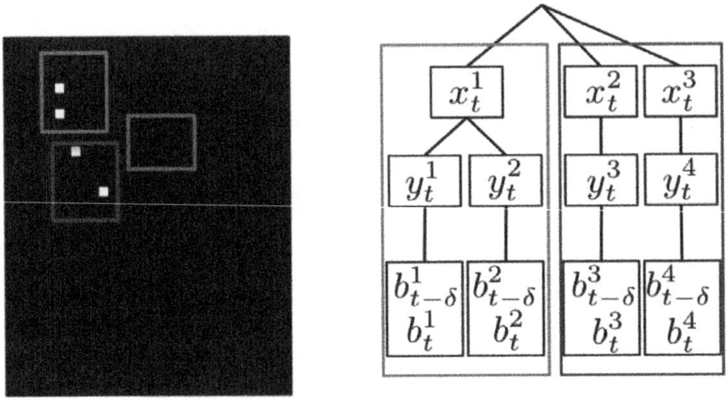

Figure 2.5. *Temporary structure suggested in [DUB 11a]. On the left, an image of the difference between $t-\delta$ and t, in which only 4 pixels changed interval, and three regions (red, green and blue) whose histogram we want to compute. On the right, the tree storing the temporal difference information, with the branches concerning the red and blue regions in red and blue (no pixel changed in the green region, so no information concerning it appears in the tree). For a color version of the figure, see www.iste.co.uk/dubuisson/tracking.zip*

We then use this temporary structure to compute a histogram from the previous one of the same region. For each pixel of the region that changed between the two images, we reduce by 1 the value contained in the interval $b_{t-\delta}$ and increase by 1 that in the interval b_t. This approach presents two primary advantages by comparison with IH. First of all, the data structure only contains the necessary information relative to the pixels changing between the images (position, previous and new interval). Furthermore, the histograms only need to be completely computed in the very first image of the sequence; it is then sufficient to update them. The general method is described in Algorithm 2.5 (we assume here that the structure coding the temporal differences has already been constructed, for more details, see [DUB 11a]).

Different properties were established (see [DUB 11a] for mathematical demonstrations) that we list below:

Algorithm 2.5: algorithm for the temporal computation of a histogram [DUB 11a]

Input : previous histogram $H_{R^i_{t-\delta}}$ of region $R^i_{t-\delta}$, temporary structure T.
Output : histogram $H_{R^i_t}$
Extract the subtree T_R of T containing the pixels of R^i_t that changed
$H_{R^i_t} \leftarrow H_{R^i_{t-\delta}}$
for each *branch* $(x) - (y) - (b_{t-\delta}, b_t)$ *of* T_R **do**
  $H_{R^i_t}(b_{t-\delta}) \leftarrow H_{R^i_{t-\delta}}(b_{t-\delta}) - 1$
  $H_{R^i_t}(b_t) \leftarrow H_{R^i_{t-\delta}}(b_t) + 1$

– the total computation time of a single histogram is always (even in the worst case) smaller for this approach than for IH. Tests showed that our approach can be up to 5,000 times faster. For example, if we consider an ROI of size 1024×1024, IH uses 7.8 ms, while the approach needs 0.03 ms if 0% of the pixels changed, and 0.5 ms if 100% of the pixels changed. The computation times are influenced very little by the quantification of histograms, as we can see in Figure 2.6: the computation times of IH increase with B (they are multiplied by 3 between $B = 16$ and $B = 256$), contrary to those of our approach, which remains stable. We can see in this same figure that this approach can be slower than IH beyond $N = 5000$. Regardless, this example was chosen to show the limits of the suggested approach, with 40% of pixels changing between the two images, which is not a very frequent case. Finally, this approach does not depend directly on the size of the ROI considered, contrary to IH (in fact, it all depends on the number of pixels changing in this ROI);

– this approach requires less memory space than IH when the histograms contain eight intervals or more. Indeed, as we have previously stated, the memory space required for IH is constant and depends on the size of the ROI and the quantification of the histograms. For this approach, it was observed that the memory space necessary becomes larger than that of the IH beyond $N = 5000$ for quantifications strictly inferior to eight. These conditions are not the most common in tracking by particle filter (for which we try not to use too many particles and use higher quantifications, in order to avoid summarizing the information too much).

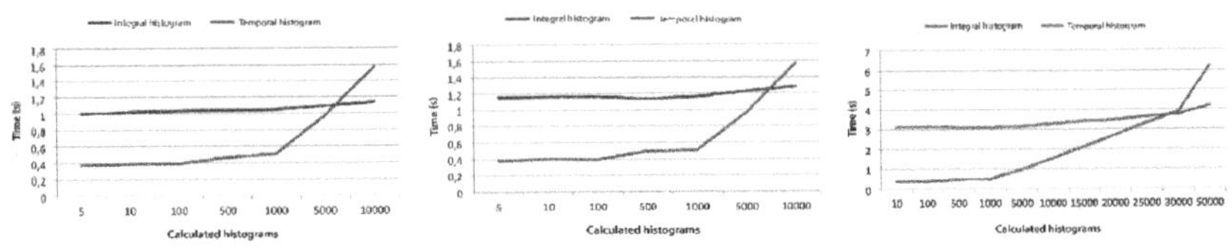

Figure 2.6. *Comparison of computation time between the IH (in blue and our approach (in red), according to the number N of histograms to compute, for a quantification, from left to right $B = \{16, 64, 256\}$, with 40% pixels changed between the two images. For a color version of the figure, see www.iste.co.uk/dubuisson/tracking.zip*

A better way to accelerate the likelihood computation is to work on the likelihood measure. A suggested algorithmic solution allowing us to accelerate the computation of the Bhattacharyya distance between two histograms, based on a reasoning similar to the one we just presented, is described below.

2.2.2.2. *Incremental distance between histograms*

There are many distances that allow us to measure the similarity between two histograms: the Bhattacharyya distance is, for example, often used to compute the distance between color histograms, although we can also use the diffusion distance [HAI 06], the L_2 norm, the *Quadratic Form Distance* [HAF 95] or even the measure of the intersection between two histograms. Temporal redundancy has been recently used to suggest a new definition, called incremental, of the Bhattacharyya distance, which has never been done previously. By using the representation of temporal changes created by the tree defined in the previous section, we propose a new definition of this distance, called the temporal Bhattacharyya distance.

DEFINITION 2.2 (Bhattacharyya distance [BHA 43]).– *Let H_{R^i} and H_{R^j} be two normalized histograms extracted, respectively, from two regions R^i and R^j of size $W \times H$. The Bhattacharyya distance $d^{\text{bt}}(H_{R^i}, H_{R^j})$ between H_{R^i} and H_{R^j} is given by [BHA 43]:*

$$d^{\text{bt}}(H_{R^i}, H_{R^j}) = \sqrt{1 - \sum_{b=0}^{B-1} \sqrt{H_{R^i}(b)H_{R^j}(b)}} \qquad [2.4]$$

Note that equation [2.4] can be written as $C(H_{R^i}, H_{R^j}) = \left(d^{\text{bt}}(H_{R^i}, H_{R^j})\right)^2 - 1$, where $C(H_{R^i}, H_{R^j}) = -\sum_{b=0}^{B-1} \sqrt{H_{R^i}(b)H_{R^j}(b)}$ is the opposite of the Bhattacharyya coefficient. As previously stated, in the case of a pixel interval change, we reduce by 1 the value of its original interval $b_{t-\delta}$ and we add 1 to its new interval b_t. If the histograms are normalized, we can subtract (respectively, add) $\frac{1}{WH}$ from (to) the value of $b_{t-\delta}$ (respectively, b_t) a constant that can be precomputed. Hence, by assuming that a single pixel changed in the region R^i between the instants $t - \delta$ and t, we can show [DUB 11a] that the Bhattacharyya coefficient is updated as follows:

$$C_t\left(H_{R^i}, H_{R^j}\right) = C_{t-\delta}\left(H_{R^i}, H_{R^j}\right) + \alpha_1 + \alpha_2 \qquad [2.5]$$

Where:

$$\alpha_1 = \left(\sqrt{H_{R^i}(b_{t-\delta})} - \sqrt{H_{R^i}(b_{t-\delta}) - \frac{1}{WH}}\right)\sqrt{H_{R^i}(b_{t-\delta})}$$

$$\alpha_2 = \left(\sqrt{H_{R^i}(b_t)} - \sqrt{H_{R^i}(b_t) + \frac{1}{WH}}\right)\sqrt{H_{R^i}(b_t)}$$

are update coefficients for each changing pixel (a trick is to use precomputed square root tables to gain time on these computations). The update corresponding to the Bhattacharyya distance is, therefore, given by:

$$\left(d_t^{\mathrm{bt}}(H_{R^i}, H_{R^j})\right)^2 = C_t\left(H_{R^i}, H_{R^j}\right) + 1 = C_{t-\delta}\left(H_{R^i}, H_{R^j}\right) + \alpha_1 + \alpha_2 + 1 \quad [2.6]$$

Hence, equation [2.6] is used for every changing pixel (in R^i or in R^j), and we can thus compute the new distance without having to compute the current histogram. The initial value $d^{\mathrm{bt}}(H_{R^i}, H_{R^j})$ is fixed at 0 as there is no temporal change in the first image and thus

$$C_0(H_{R^i}, H_{R^j}) = \left(d^{\mathrm{bt}}(H_{R^i}, H_{R^j})\right)^2 - 1 = -1.$$

Note that the implicit interest for this approach is that it is no longer necessary to compute the histograms between two time steps: only those of the regions of the first image of the sequence are necessary, as the incremental computation of the Bhattacharyya distance, described in equation [2.6], shows that only the values α_1 and α_2 need to be computed. The method is summarized in Algorithm 2.6 (here, we assume once more that the structure coding the temporal differences has already been constructed).

This computation of the incremental Bhattacharyya distance always uses less time than the computation of the classical version, and this time is directly dependent on the quantification B of the histograms, being nearly $\approx \frac{16}{23}B$ faster (whatever the number of changing pixels in the image): once $B = 2$, this approach is faster [DUB 11a]. As an example, Figure 2.7 gives a comparative study of the Bhattacharyya distance computation times between the histograms according to the ROI distance by testing several percentages of changing pixels, and we can see that even with 100%

changing pixels, our approach is faster (the scale of the ordinate axis is logarithmic, for better legibility).

Algorithm 2.6: Algorithm for computing the temporal Bhattacharyya distance between two histograms H_{R^i} and H_{R^j}, computed in the first image of the sequence [DUB 11a]

Input : the coefficient $C_{t-\delta}(H_{R^i}, H_{R^j})$, the histograms H_{R^i} and H_{R^j} computed in the first image of the sequence, the temporary structure T.

Output : the updated Bhattacharyya distance between H_{R^i} and H_{R^j}

Extract the subtree T_{R^i} of T containing the pixels of R^i that changed

$\alpha_1 \leftarrow 0$; $\alpha_2 \leftarrow 0$

for each *branch* $(x) - (y) - (b_{t-\delta}, b_t)$ *of* T_{R^i} **do**

$\quad \alpha_1 \leftarrow \alpha_1 + \left(\sqrt{H_{R^i}(b_{t-\delta})} - \sqrt{H_{R^i}(b_{t-\delta}) - \frac{1}{WH}}\right)\sqrt{H_{R^i}(b_{t-\delta})}$

$\quad \alpha_2 \leftarrow \alpha_2 + \left(\sqrt{H_{R^i}(b_t)} - \sqrt{H_{R^i}(b_t) + \frac{1}{WH}}\right)\sqrt{H_{R^i}(b_t)}$

$\left(d_t^{\text{bt}}(H_{R^i}, H_{R^j})\right)^2 = C_{t-\delta}(H_{R^i}, H_{R^j}) + \alpha_1 + \alpha_2 + 1$

Extract the subtree T_{R^j} of T containing the pixels of R^j that changed

$\alpha_1 \leftarrow 0$; $\alpha_2 \leftarrow 0$

for each *branch* $(x) - (y) - (b_{t-\delta}, b_t)$ *of* T_{R^j} **do**

$\quad \alpha_1 \leftarrow \alpha_1 + \left(\sqrt{H_{R^j}(b_{t-\delta})} - \sqrt{H_{R^j}(b_{t-\delta}) - \frac{1}{WH}}\right)\sqrt{H_{R^j}(b_{t-\delta})}$

$\quad \alpha_2 \leftarrow \alpha_2 + \left(\sqrt{H_{R^j}(b_t)} - \sqrt{H_{R^j}(b_t) + \frac{1}{WH}}\right)\sqrt{H_{R^j}(b_t)}$

$\left(d_t^{\text{bt}}(H_{R^i}, H_{R^j})\right)^2 \leftarrow \left(d_t^{\text{bt}}(H_{R^i}, H_{R^j})\right)^2 + \alpha_1 + \alpha_2 + 1$

This approach has been integrated into the particle filter and allowed a time gain of up to 75% with $N = 50$ particles and up to 20% with $N = 5000$ particles. For a larger N, our approach becomes less competitive.

In this section, we presented two approaches for accelerating the likelihood computation times when the information is represented by histograms. In the following section, we present works on information representations that are more complex and multiple, as well as solutions that we suggested for integrating them into the particle filter.

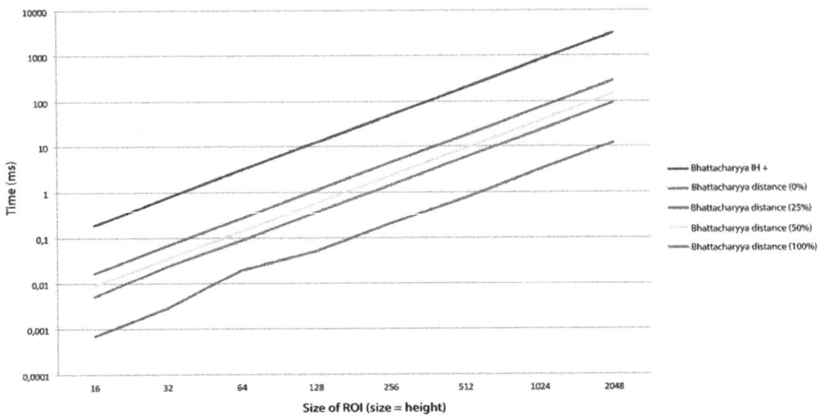

Figure 2.7. *Comparison of Bhattacharyya distance computation times between two histograms, according to the size of their region, between the method using IH, the classical distance computation (in blue) and our approach, according to the percentage of changing pixels within this region (green 100%, yellow 50%, pink 25% and red 0%). The ordinate scale is logarithmic for more readability. For a color version of the figure, see www.iste.co.uk/dubuisson/tracking.zip*

2.3. Representation of complex information

Tracking in video sequences requires considering numerous problems, among which we can mention deformation of objects over time, appearances and disappearances, changes to appearance and occlusions. We detail in this section various algorithms, all integrated into a particle filter, that provided solutions to some of these problems.

2.3.1. *Representation of observations for movement detection, appearances and disappearances*

An original contribution to detect the regions that change between two images or observations was suggested in [ABE 07b]. Based on the properties of the mutual information between two observations, it determines which regions in the new observation are moving in relation to the previous observation, and thus defines the areas containing changing measures. This model, called the *Energetic Normalized Mutual Information Model* (ENMIM) [ABE 07b], is dedicated to multiobject tracking by particle filter and allows us to manage multimodal observations. It combines a recursive partitioning method (*Quadtree Normalized Mutual Information*, QNMI), allowing it to detect ongoing measures that correspond to areas where a change was detected between two images, described below, with the energy particle filter (EPF) making associations between the measures and objects between two instants, described in more detail in Chapter 3.

The QNMI algorithm [ABE 07b] is designed for localizing regions that change between two images and which we assume corresponds to regions potentially containing a measure. It outputs a recursive quaternary partitioning (a *quadtree*) of the two images according to their statistical measure of normalized mutual information (NMI) to find their regions of similarity. The choice of NMI is justified by the fact that it allows us to take into account the different modes of acquisition of the same scene and to manage the phenomena of object appearance and/or disappearance [KNO 03]. To localize the similar regions between the two images $\mathcal{I}_t^1$ and $\mathcal{I}_t^2$, QNMI first computes their NMI measure, given by:

$$\text{NMI}(\mathcal{I}_t^1, \mathcal{I}_t^2) = \frac{E(\mathcal{I}_t^1) + E(\mathcal{I}_t^2)}{E(\mathcal{I}_t^1, \mathcal{I}_t^2)} \qquad [2.7]$$

with $E(\mathcal{I}_t^1)$ and $E(\mathcal{I}_t^2)$ are the separated entropies of the two images, measuring their complexity (the more the image is complex, the higher the value of its entropy), and $E(\mathcal{I}_t^1, \mathcal{I}_t^2)$ is their joint entropy, measuring the quantity of similar information (the more similar the images, the lower the value of their joint entropy). In this case, these entropies are computed from histograms [SHA 48]. If the two images are similar, their NMI value will be highest: we will have $\text{NMI}_{\text{max}} = 2$ if the images are not noisy and $\text{NMI}_{\text{max}} = 2 - \text{NMI}_{\text{noise}}$ otherwise; ($\text{NMI}_{\text{noise}}$ depends on the type of noise and is computed first by extracting the similar regions between the two images and computing their NMI).

If this NMI measurement is less than NMI_{max}, the images are partitioned into four subregions and the operation is repeated on the subregions of $\mathcal{I}_t^1$ and their counterparts in $\mathcal{I}_t^2$, until the minimal size T is reached in the considered regions. QNMI is summarized in Algorithm 2.7 for the comparison of two regions $B_{\mathcal{I}_t^1}$ and $B_{\mathcal{I}_t^2}$ (initially, the regions are considered to be whole images).

Algorithm 2.7: Quadtree Normalized Mutual Information (QNMI) [ABE 07b]

Function: **QNMI**$\left(B_{\mathcal{I}_t^1}, B_{\mathcal{I}_t^2}\right)$

compute the NMI$\left(B_{\mathcal{I}_t^1}, B_{\mathcal{I}_t^2}\right)$

while *NMI*$(B_{\mathcal{I}_t^1}, B_{\mathcal{I}_t^2}) < NMI_{\max}$ **and** Surface$(B_{\mathcal{I}_t^1}) > T$ **do**

  Partition $B_{\mathcal{I}_t^1}$ into 4 sub-regions $\left\{B^1_{\mathcal{I}_t^1}, B^2_{\mathcal{I}_t^1}, B^3_{\mathcal{I}_t^1}, B^4_{\mathcal{I}_t^1}\right\}$

  Partition $B_{\mathcal{I}_t^2}$ into 4 sub-regions $\left\{B^1_{\mathcal{I}_t^2}, B^2_{\mathcal{I}_t^2}, B^3_{\mathcal{I}_t^2}, B^4_{\mathcal{I}_t^2}\right\}$

  QNMI$\left(B^1_{\mathcal{I}_t^1}, B^1_{\mathcal{I}_t^2}\right)$, **QNMI**$\left(B^2_{\mathcal{I}_t^1}, B^2_{\mathcal{I}_t^2}\right)$, **QNMI**$\left(B^3_{\mathcal{I}_t^1}, B^3_{\mathcal{I}_t^2}\right)$, **QNMI**$\left(B^4_{\mathcal{I}_t^1}, B^4_{\mathcal{I}_t^2}\right)$

Figure 2.8 gives a simple example of an application between two images $\mathcal{I}_t^1$ and $\mathcal{I}_t^2$ of the `Ants` sequence. We note that a new ant appears between these two images (in the lower right corner of $\mathcal{I}_t^2$). The image on the right shows the measure detected in image $\mathcal{I}_t^2$, belonging to the region that was detected as different between the two images by the QNMI algorithm (the partitioning is indicated by red lines (see color version of the figure), the regions without detected changes are in black). Generally, every region in which changes were detected is tagged as a measure and, in the manner of the "detect-before-track" type algorithms, either associated or not with an object through a data association method, such as the method described in Chapter 3, which gave the ENMIM [ABE 07b] filter. However, this approach still creates a problem when an object has not moved between two images: its region would not be detected as a measure, and the measures–objects association will be incorrect. We, therefore, need to make the assumption that the objects remain in movement continuously throughout the sequence (in our tests, even weak movement is sufficient to detect a measure).

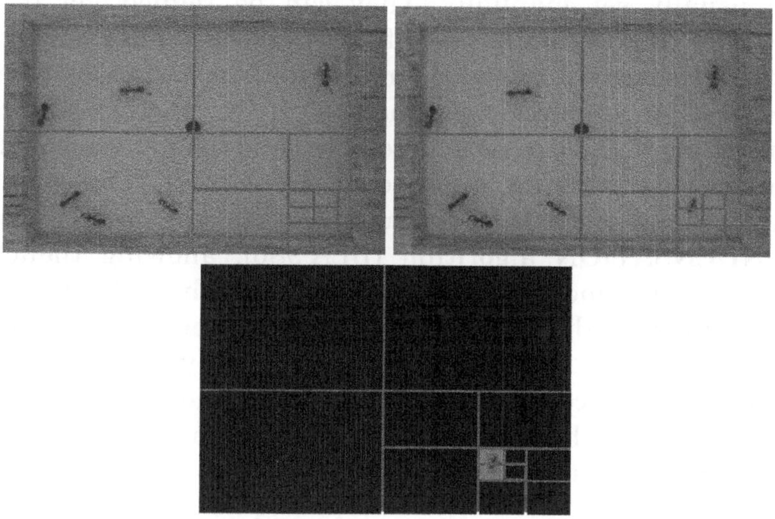

Figure 2.8. *Application of the QNMI algorithm between two images of the sequence* `Ants`[1] *with, in red lines, the obtained partitioning. On the right, the regions in black are those in which no difference was detected. For a color version of the figure, see www.iste.co.uk / dubuisson / tracking.zip*

2.3.2. *Representation of deformations*

Deformable object tracking in a video sequence is a difficult problem [COM 00, TER 92], as it implies estimating simultaneously the global movement of the object and its

1 http://www.kinetrack.org/experimental-data.html.

local deformations. A common approach consists of using a finite-dimensional representation of contours. For example, *B-Splines* were frequently used in this context and estimating the position of control points at every instant can be achieved via the CONDENSATION algorithm [BLA 98b] or the unscented Kalman filter [CHE 02, LI 04b]. According to Isard and Blake [BLA 98b, NOR 00], dynamic tracking of contours is based on the prediction and the dynamic models used. The parameters of these models to characterize realistic movement, such as constant speed and oscillations, are usually set manually. They can be refined via tests, however, this remains a difficult and complicated task, especially once we consider large spaces, in particular where we need to combine parameters. It is, therefore, more interesting to learn parameters from dynamic models for more robust tracking, adapted to the context. This is why they integrated the parameters taken from their CONDENSATION algorithm [BLA 98b], allowing them to estimate movement at each instant through particles, with the addition of the learned parameters. In practice, learning parameters has its limits for complex dynamic systems presenting non-constant speeds or non-periodic oscillations. Moreover, in the case of missing data (for instance, occlusions) learning will be disrupted.

A certain number of works use a more implicit representation of contours as sets of levels [OSH 88] for visual tracking [JAC 04, NIE 04, PAR 00, YEZ 03]. Other approaches combine *snakes* and the Kalman filter [PET 99a, PET 99b, TER 92]. In [RAT 05], others use the particle filter with active geometric contours. All of these methods are capable of following a contour that deforms and moves slowly and the parameters of the dynamic model, as those of the deformation model, are often set beforehand. Only the recent approaches that integrated active contours or *graph cuts* in particle filtering [BUG 08, RAT 05] allow us to go beyond the previously established models.

A particle filter dedicated to deformable objects (*Particle Filter for Deformable Objects*, PF-DO), suggested in [ABE 06], consists of estimating the deformations (global and local) of a contour based on its control points by working in the frequency domain. This estimation also allows us to adjust the dynamic model overtime to better follow the object. First, the contour of the object at instant $t-1$ is extracted by classical image processing methods that allow us to obtain from it a representation by a closed contour pinned along P points $\mathbf{C}_{t-1} = \{\mathbf{c}_{t-1}^1, \ldots, \mathbf{c}_{t-1}^P\}$. The first R Fourier coefficients $\{u_{t-1}^r\}_{r=1}^R$ computed over this set of points allow, after an inverse Fourier transform, to only keep a subset of R control points $\{\mathbf{p}_{t-1}^r\}_{r=1}^R$ (those located on the sharpest curves, R being fixed) from which the shape of the object is smoothed. A particle filter, whose dynamic model parameters were set at instant t according to the movement estimated at instant $t-1$, allows afterward to estimate the new positions $\{\mathbf{p}_t^r\}_{r=1}^R$ of control points. The contour is approximated by a *B-spline*, from which we extract P regularly spaced out points, to which we apply a Fourier transform to obtain R new, larger Fourier coefficients $\{u_t^r\}_{r=1}^R$. We can then compare $\{u_{t-1}^r\}_{r=1}^R$ and $\{u_t^r\}_{r=1}^R$ and, by exploiting the properties of Fourier coefficients, estimate the deformations of the curve between the instants $t-1$ and t, therefore, allowing us to update the dynamic model of the particle filter for each of the control points. Figure 2.9 gives two examples, detailed studies of which can be found in [ABE 08, ABE 06]. The sequence `Ballon` shows a balloon that, while deflating, follows a rapid, very random trajectory, as it deforms over time. Four points were used over the whole sequence (these four points being located at important curvatures, they correspond to the four extremities of the shape of the balloon, relatively oval in the image). The algorithm was also tested on the sequence `Tennis`, to show that the approach is capable of adapting to the topological changes of a contour. The black silhouette indicates the estimated shape of the curve, and the images are given solely for the purpose of comparison.

Algorithm 2.8: Algorithm of the particle filter for deformable objects (PF-DO) [ABE 06]

Input : discrete contour $\mathbf{C}_{t-1} = \{\mathbf{c}_{t-1}^1, \ldots, \mathbf{c}_{t-1}^P\}$, parameters of the dynamic
Output : discrete contour $\mathbf{C}_t = \{\mathbf{c}_t^1, \ldots, \mathbf{c}_t^P\}$, local and global deformations
$\{u_{t-1}^r\}_{r=1}^R \leftarrow R$ first Fourier coefficients of the discrete contour $\mathbf{c}_{t-1}$
$\{\mathbf{p}_{t-1}^r\}_{r=1}^R \leftarrow$ main points of the smoothed contour, obtained by inverse Fourier transform
for $r = 1$ **at** R **do**
  Estimate the position of $\mathbf{p}_t^r$ with a SIR filter (see chapter 1), according to the current dynamic
Reconstruct $\mathbf{C}_t$ from $\mathbf{p}_t^r$ with P points
$\{u_t^r\}_{r=1}^R \leftarrow R$ first Fourier coefficients of the discrete contour $\mathbf{C}_t$
Comparer $\{u_{t-1}^r\}_{r=1}^R$ et $\{u_t^r\}_{r=1}^R$ to determine the local and global transformations of the curve between $t-1$ and t
Update the parameters of the dynamic [ABE 08]

2.3.3. *Multifeature representation*

Integrating several features into the tracking allows us to use complementary observations in the case where some of them are not informative enough at a given instant. Thus, the object to track is defined by a set of features (color, gradient, shapes, etc.), which need to be fused [PÉR 04, TRI 01, WU 01]. Needless to say, it is necessary for the features to be orthogonal [NIC 08] (i.e. they do not supply informations derived from each other), if we wish, in the case where one of them is not discriminatory enough at a certain instant, to use another to characterize the object correctly.

The literature is extremely abundant, however, the most general method is to fuse the features [BIR 98]. In this case, every feature supplies a likelihood or a matching score for the possible positions of the object and the final output is the product of these scores or likelihoods. The main problem with this type of approach is that the same significance is attributed to each of the features, which is not realistic and often leads to estimation errors. For example, in [BAD 07, MAG 05b], it corresponds to the average of the sample dispersion related to a feature and, in [COL 05], it is seen as the success measure to characterize the object with respect to the background after estimation. We can group the approaches that use several features and their confidence into two groups.

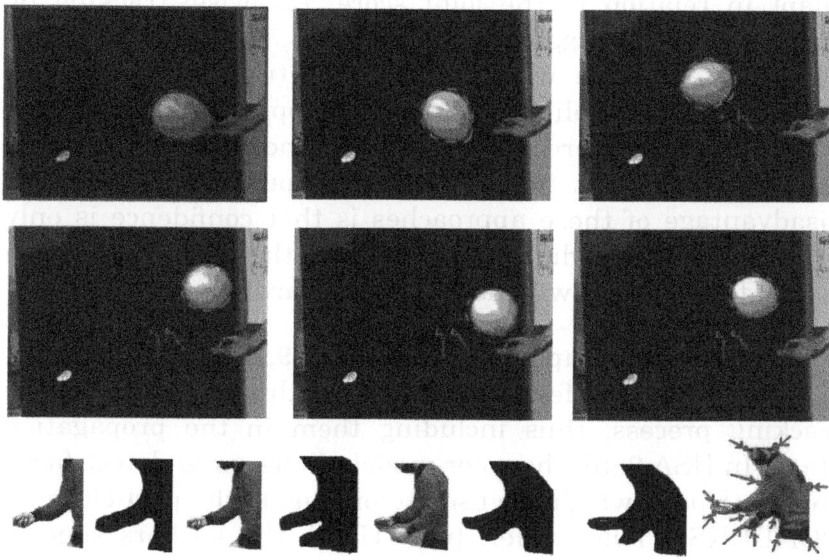

Figure 2.9. *Two examples of tracking deformable objects obtained by the PF-DO method [ABE 06]. In the first two rows, the* `Ballon` *sequence (see [ABE 08] for a more complete study of tracking) and, in the last row, the* `Tennis` *sequence[2]: the black shapes are the silhouettes obtained from the estimated contours and the color images are given for purposes of comparison. The last image shows the position of the considered control points. For a color version of the figure, see www.iste.co.uk/dubuisson/tracking.zip*

The first group [BRA 07a, MAG 05b, SHE 03, SPE 03, TRI 01] attributes each feature a confidence and uses it during the correction stage. In [SPE 03, TRI 01], we compute

2 http://i21www.ira.uka.de/image_sequences/.

the weighted average of the confidence maps of each feature, where each confidence is relative to the success of the others. Hence, the confidences are computed by considering the correlations between visual features. This means that the confidence of a feature is defined according to the success of other features used for tracking. The features are looking to reach some sort of global agreement between themselves and adapt in relation to the joint score. Likewise, tracking by particle filter as suggested in [BRA 07a, MAG 05b, SHE 03] attributes adaptive weights to the features to estimate a combination of likelihoods. In these approaches, the global likelihood is more precise as the confidence of each feature is taken into account in the global computation. The major disadvantage of these approaches is that confidence is only taken into account during correction: if the prior assumption is bad, the tracking will also be bad, regardless.

The second group [CHE 04, DUF 09, ISA 98b, PÉR 04, WU 01] makes the features interact with the entirety of the tracking precess, thus including them in the propagation stage. In [ISA 98b], the color model allows us to detect fuzzy shapes (*blobs*), which are used to propagate the particles in a CONDENSATION model. In [WU 01], the authors suggest estimating the coinference of the features by separating the space along two features (shape and color) and using one to supply the other with samples (hypotheses) and vice versa. Hence, in their formulation, the shapes sample is propagated according to the measurements associated with colors and vice versa. A sampling partitioned into two stages is suggested in [PÉR 04], where the first stage uses two features (sound and movement) to obtain a rough estimation of the object. The second stage refines the estimation by, this time, using only the color. Two stages are also suggested in [CHE 04], however, in this case, all the features are first tracked independently, then fused to construct better hypotheses. The originality lies in the feedback sent to each individual tracking to improve it, in relation to this refined estimation. Finally, in [DUF 09], the order in which the features are taken into account in the process is changed dynamically, according to their confidence.

In what follows, we detail the contributions, integrated into a particle filter, for multifeature tracking by taking into account the confidences attributed to each one and specifically, by evolving them over time. The first approach integrates the properties of two families of approaches developed previously and suggests multimodal tracking [ERD 12b]. The second approach is a multifragment tracking solution without a constraint model or learning [ERD 12a]. Here, the confidences associated with fragments to suggest a more robust joint estimation. They are the topic of the upcoming subsections.

2.3.3.1. *Multimodal tracking*

In this approach, a tracking algorithm by particle filter integrates multimodal information at both the prediction and correction stages. Similar works were suggested in [NIC 08], which also combine democratic integration [TRI 01] with two-staged sampling [PÉR 04]. The difference is that in the approach presented here, the structure is not fixed: every layer is predefined and independent of the other. This allows interactions between the different features, which make the approach more flexible and adaptive over time.

During the prediction stage, every particle is associated with a particular feature and propagated according to a proposal function related to this feature. One of the originalities of this approach is that this association can evolve over time, i.e. a particle can be attached to a different feature based on the instants considered and this association depends on the confidence of each feature: the higher the confidence of a feature, the higher the number of particles associated with it and vice versa. Hence, contrary to other approaches [PÉR 04, WU 01], the number of hypotheses associated with each feature is not fixed and can vary over time (even if the overall sum always remains equal to the total number N of particles). This dynamic process improves significantly the prediction process, as it is adaptive.

For each feature, we define a proposal function and a likelihood function.

Color features. By following the works suggested in [PÉR 02], the color observation model is a color histogram in the ***Hue-Saturation-Value*** (HSV) space, with $B_C = B_h B_s + B_v$ bins. The likelihood is thus defined by:

$$p(\mathbf{y}_t^C \mid \mathbf{x}_t^{(i)}) \propto \exp\left(-\frac{D^2(\mathbf{h}_{\mathbf{x}}^C, \mathbf{h}_{ref}^C)}{2\sigma_C^2}\right) \quad [2.8]$$

where $\mathbf{h}_{ref}^C$ is the normalized histogram containing B_C bins, $\mathbf{h}_{\mathbf{x}}^C$ is the normalized histogram of the candidate region associated with a particle $\mathbf{x}_t^{(i)}$ and $D^2(\mathbf{h}_{\mathbf{x}}^C, \mathbf{h}_{ref}^C)$ is the squared Bhattacharyya distance [BHA 43] between two histograms. Finally, σ_C is the standard deviation associated with the Gaussian. The proposal function is defined as follows:

$$q^C(\mathbf{x}_t, \mathbf{y}_t \mid \mathbf{x}_{t-1}, \mathbf{y}_{t-1}, \mathbf{y}_t^C) = \beta_{RW}\mathcal{N}\left((\mathbf{x}_{t-1}, \mathbf{y}_{t-1}), (\sigma_x^2, \sigma_y^2)\right) + \frac{(1-\beta_{RW})}{N_C}\sum_{i=1}^{N_C}\mathcal{N}\left(\mathbf{p}_i^C, (\sigma_x^2, \sigma_y^2)\right) \quad [2.9]$$

In equation [2.9], the first part is a random Gaussian step centered on the previous estimation of the object's position. The points $\mathbf{p}_i^C = (x_i, y_i)$, $i = 1, \cdots, N_C$ are the positions of the image where likelihood is above a given threshold (i.e. $p(\mathbf{y}_t^C \mid \mathbf{x}_t) > \tau^C$), on which we centered the Gaussian steps. The second part is the equation allowing us to jump from the state space to plausible positions of the objects at the current instant. The β_{RW}, σ_x and σ_y parameters need to be set.

Movement feature. The movement present between two images at the instants $t-1$ and t can be discovered by computing their absolute difference, in which pixels with a higher grayscale value correspond to movement (or noise). Here again, the approach followed in [PÉR 02] is used to determine likelihood. To the interest region, associated with a state $\mathbf{x}_t$, we associate a histogram of the movement M $\mathbf{h}^M_{\mathbf{x}_t} = (h^M_{1,\mathbf{x}_t}, \cdots, h^M_{B_M,\mathbf{x}_t})$ with B_M the number of bins. The reference histogram $\mathbf{h}^M_{ref}$ is defined by a uniform distribution, i.e. $h^M_{i,ref} = 1/B_M$, $i = 1, \cdots, B_M$. When there is no movement, the Bhattacharyya distance between the two histograms is given by $D^2_{no\_mot.} = 1 - \sqrt{1/B_M}$. Hence, the likelihood of movement for the particle $\mathbf{x}^{(i)}_t$ is given by:

$$p(\mathbf{y}^M_t \mid \mathbf{x}^{(i)}_t) \propto 1 - \exp\left(-\frac{D^2_{no\_mot.} - D^2(\mathbf{h}^M_{\mathbf{x}_t}, \mathbf{h}^M_{ref})}{2\sigma^2_M}\right) \quad [2.10]$$

Along the same principle as for the proposal function associated with a color, we define a proposal function associated with movement by:

$$q^M(\mathbf{x}_t, \mathbf{y}_t \mid \mathbf{x}_{t-1}, \mathbf{y}_{t-1}, \mathbf{y}^M_t) = \beta_{RW}\mathcal{N}\left((\mathbf{x}_{t-1}, \mathbf{y}_{t-1}), (\sigma^2_x, \sigma^2_y)\right) + \frac{(1-\beta_{RW})}{N_M}\sum_{i=1}^{N_M}\mathcal{N}\left(\mathbf{p}^M_i, (\sigma^2_x, \sigma^2_y)\right). \quad [2.11]$$

The β_{RW}, σ_x and σ_y parameters need to be set. The points $\mathbf{p}^M_i = (x_i, y_i)$, $i = 1, \cdots, N_M$ are the positions of the image where the likelihood is above a given threshold.

Infrared feature. Infrared gives an idea of the position of moving objects, and the likelihood is defined by:

$$p(\mathbf{y}^I_t \mid \mathbf{x}^{(i)}_t) \propto \exp\left(-\frac{D^2(\mathbf{h}^I_{\mathbf{x}_t}, \mathbf{h}^I_{ref})}{2\sigma^2_I}\right) \quad [2.12]$$

where $\mathbf{h}^I_{ref} = (h^I_{1,ref}, \cdots, h^I_{B_I,ref})$ is the normalized reference histogram of B_I bins and $\mathbf{h}^I_{\mathbf{x}_t} = (h^I_{1,\mathbf{x}_t}, \cdots, h^I_{B_I,\mathbf{x}_t})$ is that of the candidate region. The proposal function is given by:

$$q^I(\mathbf{x}_t, \mathbf{y}_t \mid \mathbf{x}_{t-1}, \mathbf{y}_{t-1}, \mathbf{y}^I_t) = \beta_{RW}\mathcal{N}\left((\mathbf{x}_{k-1}, \mathbf{y}_{k-1}), (\sigma^2_x, \sigma^2_y)\right) + \frac{(1-\beta_{RW})}{N_I}\sum_{i=1}^{N_I}\mathcal{N}\left(\mathbf{p}^I_i, (\sigma^2_x, \sigma^2_y)\right) \quad [2.13]$$

where $\mathbf{p}^I_i = (x_i, y_i), i = 1, \cdots, N_I$ are points of the image whose value is above a given threshold.

The global likelihood is defined by using the measures for each feature, as well as their confidence. This allows us to have a more precise likelihood in the correction stage, and hence the disruptive effects of low confidence are erased.

At the initialization stage, the particles are generated randomly around the position $\mathbf{x}_0$ (defined manually) of the object to follow, according to a uniform law ($p(\mathbf{x}_0) = \mathcal{U}_{\mathcal{X}}(\mathbf{x}_0)$). The confidences attributed to each feature are equiprobable, let $r^l_0 = 1/L$, where $l = 1, \ldots, L$ is the feature index. At every instance t, the method is divided into three main steps, which we describe briefly here (see [ERD 12b] for more details) and that are described in Algorithm 2.9:

1) Adjusting confidences r^l_t. By taking the definition of democratic integration [TRI 01], the confidences are updated (equation [2.16]) by considering both the global success of the features l in the past (given by r^l_{t-1}) and the approximated target position $\hat{\mathbf{x}}_t$ (equation [2.16]). Hence, a feature whose quality increases will see its confidence increase over the next steps.

2) Generalizing new hypotheses. The idea is that the higher the confidence of a feature l, the higher the number of the particles representing it in the tracking. To this

end, we generate new hypothesis according to a random multinomial selection of confidences (equation [2.17]). The particles associated with a feature are then propagated via the proposal function $q_t^l(.)$ of this feature (these functions are described in [ERD 12b] and are a combination of a random Gaussian step around the previous estimated position of the object and the Gaussian steps centered around feature points of the image of the corresponding feature, defined by a detection process).

3) Updating weights of the particles. We use here the classical scheme (equation [2.14]). The critical point is that the global likelihood is also defined adaptively, according to the confidence (equation [2.18]).

This approach proved to be very efficient in comparison to sampling partitioned into two layers [PÉR 04] and dynamically partitioned sampling (DPS) [DUF 09], as we can see from the example in Figure 2.10, where color and movement feature were used. Our approach (Figure 2.10(a)) removes all ambiguities, in particular when a group hides the tracked person: the confidence of the color decreases and that of the movement increases. Conversely, when the tracked person does not move anymore, it is movement confidence that decreases. The close-ups in Figure 2.11 illustrate this phenomenon: very few blue particles (associated with movement) are generated once the person stops moving, while the number of green particles (associated with color) remains important. Figure 2.10(b) shows the disadvantage of the method suggested in [PÉR 04], which works first on movement and then on color (the particles for this case are represented in yellow, as they are not attached to any feature): the tracking is disrupted when another person enters the scene (image 4). In this example, our approach gives similar results to DSP, which considers a specific processing order of features based on their confidence.

Algorithm 2.9: Algorithm of multi-modal tracking suggested in [ERD 12b]

Input : set of particles $\{\mathbf{x}_{t-1}^{(i)}, w_{t-1}^{(i)}\}_{i=1}^{N}$

Output : set of particles $\{\mathbf{x}_{t}^{(i)}, w_{t}^{(i)}\}_{i=1}^{N}$

computation of the confidence $\{r_t^l\}$ of the features according to the current observation $\mathbf{y}_t$

– Approximating the position of the object $\hat{\mathbf{x}}_t$ with the previous confidences and the current observation:

$$\hat{\mathbf{x}}_t = \arg\max_x (p(\mathbf{y}_t \mid \mathbf{x}_t)) = \arg\max_x \left(\prod_{l=1}^{L} p(\mathbf{y}_t^l \mid \mathbf{x}_t)^{r_{t-1}^l} \right) \quad [2.14]$$

– Estimating the quality of each feature:

$$\bar{s}_t^l = \begin{cases} 0 & \text{si } p(\mathbf{y}_t^l \mid \hat{\mathbf{x}}_t) \leq \frac{1}{L}\sum_{k=1}^{L} p(\mathbf{y}_t^k \mid \mathbf{x}_t) \\ p(\mathbf{y}_t^l \mid \hat{\mathbf{x}}_t) - \frac{1}{L}\sum_{k=1}^{L} p(\mathbf{y}_t^k \mid \mathbf{x}_t) & \text{otherwise} \end{cases} \quad [2.15]$$

– computing normalized qualities $s_t^l = \bar{s}_t^l / \sum_j \bar{s}_t^j$

– Updating the confidences with current normalized qualities:

$$r_t^l = r_{t-1}^l + \eta(s_t^l - r_{t-1}^l), \eta \in [0, 1] \quad [2.16]$$

$\left\{\mathbf{x}_{t-1}^{(i)}, 1/N\right\}_{i=1}^{N} \leftarrow$ multinomial resampling of $\left\{\mathbf{x}_{t-1}^{(i)}, w_{t-1}^{(i)}\right\}_{i=1}^{N}$

Propagation: generation of new hypotheses $\{\mathbf{x}_t^{(i)}\}_{i=1}^{N}$

$$\text{– Set } l_t^{(i)} = \begin{cases} 1 & \text{si } \alpha^{(i)} < r_t^1 \\ \dots & \\ l & \text{si } \sum_{j=1}^{l-1} r_t^j \leq \alpha^{(i)} < \sum_{j=1}^{l} r_t^j \quad \text{with } \alpha^{(i)} \sim \mathcal{U}(0,1) \\ \dots & \\ L & \text{si } \alpha^{(i)} \geq \sum_{j=1}^{L-1} r_t^j \end{cases} \quad [2.17]$$

– Simulation $\mathbf{x}_t^{(i)} \sim q^{l_t^{(i)}}(\mathbf{x}_t \mid \mathbf{x}_{t-1}^{(i)}, \mathbf{y}_t^{l_t^{(i)}})$

Correction: computation of new weights $\{w_t^{(i)}\}_{i=1}^{N}$

– computing global likelihood:

$$p(\mathbf{y}_t \mid \mathbf{x}_t^{(i)}) = \prod_{l=1}^{L} p(\mathbf{y}_t^l \mid \mathbf{x}_t^{(i)})^{r_t^l} \quad [2.18]$$

– Updating weights:

$$w_t^{(i)} \propto w_{t-1}^{(i)} \frac{p(\mathbf{y}_t \mid \mathbf{x}_t^{(i)}) p(\mathbf{x}_t^{(i)} \mid \mathbf{x}_{t-1}^{(i)})}{q^{l_t^{(i)}}(\mathbf{x}_t^{(i)} \mid \mathbf{x}_{t-1}^{(i)}, \mathbf{y}_t^{l_t^{(i)}})} \quad [2.19]$$

– Normalizing: $\left\{w_t^{(i)}\right\}_{i=1}^{N} \leftarrow \left\{\frac{w_t^{(i)}}{\sum_{j=1}^{N} w_t^{(j)}}\right\}_{i=1}^{N}$

a) b) c)

Figure 2.10. *Sequence issued from the base* BEHAVE[3]. *Several results of the tracking (close-ups) with* $N = 100$ *obtained by: a) the proposed approach with confidences set according to the context, b) the two-layer partitioned sampling method [PÉR 04] and c) the DSP method [DUF 09]. The white rectangles are estimations, the green rectangles are the best particles associated with color, the blue rectangles are those associated with movement and the yellow rectangles are not associated with any modality (case of the approach described in [PÉR 04]). For a color version of the figure, see www.iste.co.uk / dubuisson / tracking.zip*

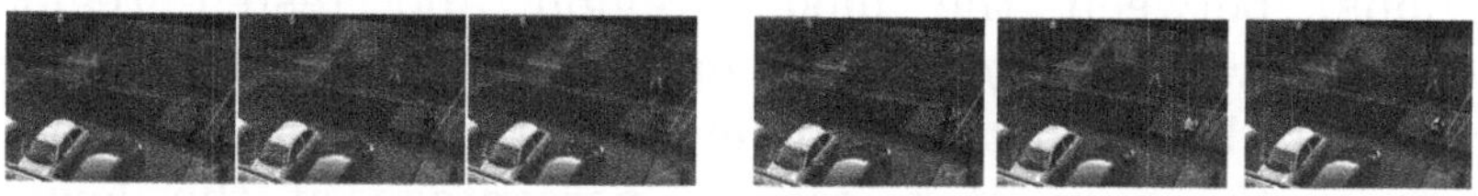

$r_{260}^{\text{coul}} = 0.71$ $r_{500}^{\text{coul}} = 0.99$ $r_{526}^{\text{coul}} = 0.94$ $r_{260}^{\text{mouv}} = 0.3$ $r_{500}^{\text{mouv}} = 0.01$ $r_{526}^{\text{mouv}} = 0.06$

Figure 2.11. *Sequence issued from the base* BEHAVE[3] *(close-ups on images 260, 500 and 526). Visual representation of confidences associated with color (green) and movement (blue) features. The higher the confidence, the larger the number of particles (here, the green and blue points) forming the sample that represents it. For a color version of the figure, see www.iste.co.uk / dubuisson / tracking.zip*

2.3.3.2. *Multifragment tracking*

Here, we are interested in explicitly modeling an object by a set of fragments (that are seen as features of the object). For example, we can model the human body with a set of

3 http://groups.inf.ed.ac.uk/vision/BEHAVEDATA/INTERACTIONS/.

fragments corresponding to the head, torso, arms, etc. and to consider implicitly the spatial arrangement between the fragments, which allows us to track articulated and non-rigid objects [CHO 09, NEJ 08, NIC 08] more efficiently. However, in this case, we use explicit object representation models (i.e. every part concerns a particular part of the object, for example, the head or an arm), which make these models closely bound to the application.

Works attempted to break free from the requirement of prior knowledge of the object model (Figure 2.12 gives an example of the difference between these two types of model for a face). For example, the model *FragTrack* [ADA 06] describes the object by a region and its detection in the image $\mathcal{I}_t$ is accomplished through matching. The originality is that the region is divided into a set of arbitrary fragments, each one describing a section of the object and participating in the global detection. The idea then is to evaluate the possible positions (x, y) for the object at instant t by measuring the dissimilarity (EMD distance between the histograms of the regions) between the model region and tested region, centered around (x, y). To this end, a score for each couple of fragments between two regions is evaluated and the global score is the sum of a certain percentage of the lowest obtained scores (in their case, 25%). The new position of the region will be the one with the smallest score.

Among the other approaches that do not use an explicit model, there is also a series of works [BAB 09, GRA 06a, GRA 08, WOO 07] that use approaches without prior model, but include a *boosting* phase, that is online learning via a set of simple classifiers, often Haar-like features, combined to obtain a more robust classifier. Due to their learning, these algorithms are today among the best for quick and precise tracking.

There is a tracking algorithm [ERD 12a] that does not use any explicit learning model, and the global method is close enough to the method described in Algorithm 2.9, with a few differences. The state to estimate $\mathbf{x}_t$ is a region defined by its center and the size of a particle $\mathbf{x}_t^{(i)}$ corresponds to a possible region that can be decomposed into fragments. The likelihoods are linked not to modalities, but to fragments (and thus here, L is the number of fragments and $\mathbf{y}_t^l$ is the measure for the fragment l, s_t^l is its quality and r_t^l is its confidence). Finally, during the prediction stage, all the particles are propagated according to the object transition function (Gaussian centered on the previous estimated position): therefore, we use the *Bootstrap* filtering principle [ISA 98a].

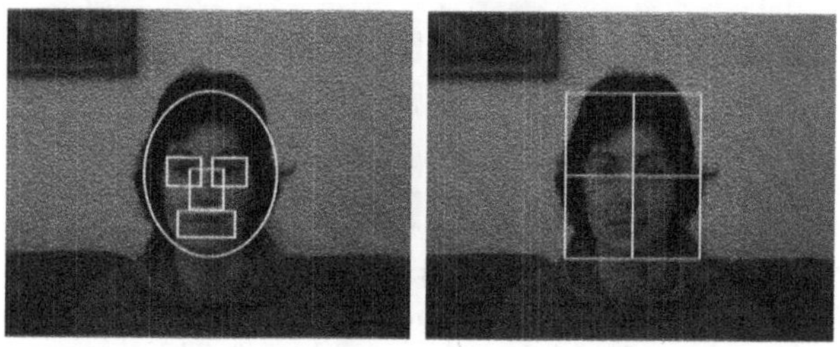

Figure 2.12. *Difference between an explicit model (on the left) and a free model (on the right)*

This approach is inspired by *FragTrack* [ADA 06], but improves the performance of several aspects, which are described below:

– one of the downsides of *FragTrack* is that it is necessary to make an assumption concerning the minimum size of the object's visible part, corresponding to a percentage of the fragments, which can create problems if this value is set incorrectly. To solve this issue, we use an adaptive feature integration scheme [TRI 01]. The confidence factor, dependent on the context, is associated with every fragment that represents a part of the object and contributes, according to its reliably, to joint tracking. Thus, an occluded fragment, which will have low confidence, will not be taken into account by the tracking, which makes our algorithm robust to partial occlusions and to changes in appearance, without resorting to learning. This is illustrated in Figure 2.15, in which we see that the tracking of the person is done correctly when we use adaptive confidences (bottom row), even in the case of partial occlusion;

– the adaptive integration scheme allows us to know at every instant which parts of the tracked object are visible or hidden, as illustrated in Figure 2.13: the fragments with higher confidences (the least transparent) are those that are visible at the considered moment and vice versa;

Figure 2.13. *Sequence* `Occluded Face`[4]. *Tracking obtained with the approach from [ERD 12a] superimposed on confidence maps (the less transparent the green, the higher the confidence). This approach makes it possible to manage partial occlusions, and the value of the confidence allows us to determine which fragments are occluded. For a color version of the figure, see www.iste.co.uk/dubuisson/tracking.zip*

– tracking is done by particle filter, which makes it possible not to exhaustively search the state space, making the suggested approach more efficient in terms of computation time. Moreover, the particle filter allows us to integrate other parameters naturally, such as the size and orientation of the windows enclosing the object;

– adaptive formulation makes it possible to have more precise and robust tracking than *FragTrack* and supersedes some algorithms that use online learning, such as *MILTrack* [BAB 09], as shown in Figure 2.14.

4 http://vision.ucsd.edu/ bbabenko/project_miltrack.shtml.

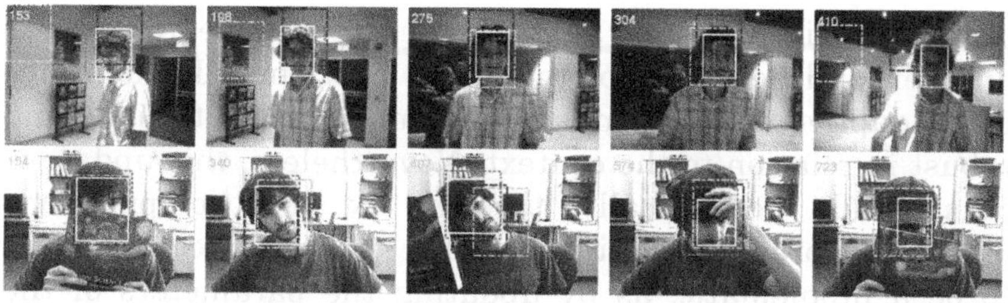

Figure 2.14. *Tracking obtained for the sequences* `David Indoor`[5] *(first row) and* `Occluded Face 2`[5] *(second row) for different approaches:* MILTrack *(blue),* FragTrack *(green), the approach from [ERD 12a] with a fixed-size window (pink), the approach from [ERD 12a] with a variable size window (yellow). For a color version of the figure, see www.iste.co.uk / dubuisson / tracking.zip*

Figure 2.15. *Some examples of the tracking result for the sequence* `Woman`[6]. *In the first row, the reliabilities are fixed and in the second row, they evolve over time as a function of the context. For a color version of the figure, see www.iste.co.uk / dubuisson / tracking.zip*

2.3.3.3. *Multiappearance tracking*

Here, we attempt to integrate several appearances of the same object into the particle filter. In general, the integration of a model with a changing appearance consists of progressively updating the reference model. In [MUN 08a, NUM 03a], the reference histogram is updated at every instant as a function of its previous state and new estimate, if it has a high enough likelihood. Following our discussion in sections 2.3.3.1 and 2.3.3.2, we can define

5 http://vision.ucsd.edu/ bbabenko/project_miltrack.shtml.

6 http://www.cs.technion.ac.il/ amita/fragtrack/fragtrack.htm.

confidences in a way so as to privilege one appearance over another, and hence adapt the model over time. Incidentally, this adaptive feature makes the model both flexible and robust in relation to the context. Nevertheless, this update is often made heuristically, through ex post computations following a predefined criterion. This specifically can turn out to be a problematic, as by updating the parameters of the model, the likelihood thus considered becomes obsolete, which has a high chance of penalizing the tracking algorithm by diverging it.

The model suggested in [WID 11a] defines, as previously, the likelihood by the probability density of a mixture, whose weights are attached to particles. The originality lies in the fact that a weight is linked, not to a feature or a sensor, but to a state or observation decomposition. An appearance is defined explicitly via a feature of the object that can be different from that used in the likelihood. The suggested approach is summarized in Figure 2.16, and we detail it below.

Let $\mathbf{a} = \{\mathbf{a}^1, \ldots, \mathbf{a}^R\}$ be a vector containing R appearances, where $\mathbf{a}^r$ is the appearance related to the posture r. For example $\mathbf{a}^1$ represents the appearance of the posture 1, *smile*, of a mouth. The joint likelihood is written as a decomposition:

$$\begin{aligned} p(\mathbf{y}_t|\mathbf{x}_t, \mathbf{a}) &= \sum_{r=1}^{R} p(\mathbf{y}_t, \mathbf{l} = r|\mathbf{x}_t, \mathbf{a}^r) \\ &= \sum_{r=1}^{R} \varphi^r\left(\gamma(\mathbf{x}_t, \mathbf{y}_t)\right) p(\mathbf{y}_t|\mathbf{x}_t, \mathbf{a}^r) \end{aligned} \quad [2.14]$$

where $\mathbf{l}$ designates a posture, $p(\mathbf{y}_t|\mathbf{x}_t, \mathbf{a}^r)$ is the likelihood of the r and $\varphi^r\left(\gamma(\mathbf{x}_t, \mathbf{y}_t)\right)$ is the weight of the posture r associated with the feature $\gamma(\mathbf{x}_t, \mathbf{y}_t)$ (see [WID 10a, WID 11a]). In our case, the observation vector is decomposed into two parts: a posture component $\mathbf{y}_t^{\text{Pos}}$ used for the computation of the likelihood and an appearance component $\mathbf{y}_t^{\text{App}}$ used for the computation of the appearance probabilities. By assuming l conditionally independent given $\mathbf{y}_t^{\text{App}}$ given $\mathbf{y}_t^{\text{Pos}}$, the likelihood with multiple models is written as:

$$p(\mathbf{y}_t|\mathbf{x}_t, \mathbf{a}) \approx \sum_{r=1}^{R} \varphi^r \left(\gamma(\mathbf{x}_t, \mathbf{y}_t^{\text{Pos}})\right) p(\mathbf{y}_t^{\text{App}}|\mathbf{x}_t, \mathbf{a}^r) \qquad [2.15]$$

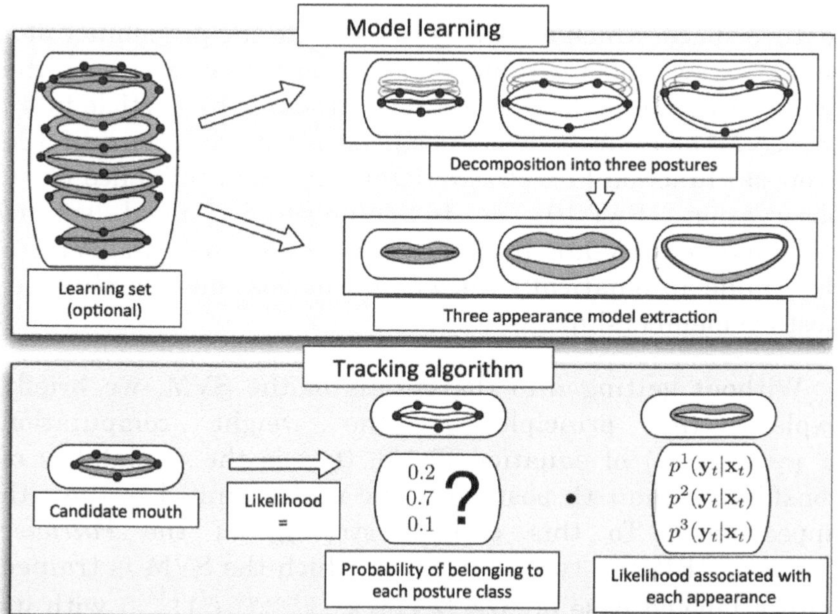

Figure 2.16. *Illustration of the suggested approach (derived from [WID 10a]). Learning the models is accomplished by support vector machine (SVM), which decomposes the object, here a mouth, into several classes, which we call postures. In this example, blue control points are used to define the SVM separation criterion. The learning base, as well as the SVM classifier, allows us to determine three appearance models, which, in this example, use color information (extraction of the three appearance models). In the tracking algorithm, when a candidate mouth is suggested, the total likelihood corresponds to the weighting of the marginal likelihoods in each appearance with the learning weights of the shape at the considered postures. For color version of the figure, see www.iste.co.uk/dubuisson/tracking.zip*

This approximation is used to identify the studied posture and appearance decomposition. The original feature of the decomposition made in equation [2.15] is the conditioning of the probability of l in relation to $\mathbf{x}_t$ and $\mathbf{y}_t^{\mathrm{Pos}}$. This means that the weights $\varphi^r\left(\gamma(\mathbf{x}_t, \mathbf{y}_t^{\mathrm{Pos}})\right)$ are determined with the last observation and for the considered particle, which differs from the methods of the literature.

The methodology consists of using a supervised classifier, here a multiclass and probabilistic SVM [BIS 07] in the feature space, which allows us to compute the probability of a couple $(\mathbf{x}_t, \mathbf{y}_t^{\mathrm{Pos}})$ being a member of a posture class r with the feature function γ. In the context of tracking by particle filter, the SVM has rarely been used: either as detectors of the proposal function [YE 09], to determine the confidence of an observation [PAT 10] or to estimate its ideal set of parameters [LIA 09]. In this approach, we use an SVM to determine automatically the weights of class membership, i.e. posture membership.

Without getting into the details of the SVM, we briefly explain the principle of the weight computation $\varphi^d\left(\gamma(\mathbf{x}_t, \mathbf{y}_t^{\mathrm{Pos}})\right)$ of equation [2.15], that is the probability of considering the r-th posture of the model (linked to the r-th appearance). To this end, a strategy of the *Pairwise Coupling* [BIS 07] type is used, in which the SVM is trained with a labeled base of size D $\left\{\left(\gamma(\mathbf{x}^d, \mathbf{y}^{\mathrm{Pos},d}), l^d\right)\right\}_{d=1}^{D}$, with its feature function γ (which corresponds, for example, to a color histogram or a set of control points defining a shape) and the tag l^d, which indicates the index of the considered posture and thus designates implicitly the appearance used. The weights $\varphi^r\left(\gamma(\mathbf{x}_t, \mathbf{y}_t^{\mathrm{Pos}})\right)$ are then simply obtained by computing the probabilities $\{p_r\}_{r=1}^{R}$ such that

$\forall r \in \{1, \ldots, R\}, \varphi^r(\gamma(\mathbf{x}_t, \mathbf{y}_t^{\text{Pos}})) = \mathbb{P}(1 = r|\gamma(\mathbf{x}_t, \mathbf{y}_t^{\text{Pos}})) = p_r$. The appearance models are then defined by the following formula:

$$\mathbf{a}^r = \frac{\sum_{d=1}^{D} \mathbb{P}(1 = r|\gamma(\mathbf{x}^d, \mathbf{y}_t^{\text{Pos}}))\eta(\mathbf{x}^d)}{\sum_{d=1}^{D} \mathbb{P}\left(1 = r|\gamma(\mathbf{x}^d, \mathbf{y}^{\text{Pos},d})\right)} \qquad [2.16]$$

where $\mathbb{P}\left(1 = r|\gamma(\mathbf{x}^d, \mathbf{y}^{\text{Pos},d})\right)$ is the probability of considering the posture r in relation to the feature $\gamma(\mathbf{x}^d, \mathbf{y}^{\text{Pos},d})$ and η is the function characterizing an appearance (see [WID 10a, WID 11a] for more details).

The particle filter procedure for the integration of multiple appearance models is presented in Algorithm 2.10.

Algorithm 2.10: Particle filter for the integration of multiple appearance models [WID 11a]

Input : set of particles $\{\mathbf{x}_{t-1}^{(i)}, w_{t-1}^{(i)}\}_{i=1}^N$, probabilities $\{p_r\}_{r=1}^R$, appearances $\{\mathbf{a}^r\}_{r=1}^R$

Output : set of particles $\{\mathbf{x}_t^{(i)}, w_t^{(i)}\}_{i=1}^N$

$\left\{\mathbf{x}_{t-1}^{(i)}, 1/N\right\}_{i=1}^N \leftarrow$ multinomial resampling of $\left\{\mathbf{x}_{t-1}^{(i)}, w_{t-1}^{(i)}\right\}_{i=1}^N$

for $i = 1$ **at** N **do**

Propagation: $\mathbf{x}_t^{(i)} \sim q(\mathbf{x}_t|\mathbf{x}_{t-1}^{(i)}, \mathbf{y}_t)$

Compute $\varphi^r\left(\gamma(\mathbf{x}_t^{(i)}, \mathbf{y}_t^{\text{Pos}})\right) = \mathbb{P}(1 = r|\gamma(\mathbf{x}_t^{(i)}, \mathbf{y}_t^{\text{Pos}}))$

Correction:

$$w_t^{(i)} = w_{t-1}^{(i)} \frac{\left[\sum_{r=1}^R \varphi^r\left(\gamma(\mathbf{x}_t^{(i)}, \mathbf{y}_t^{\text{Pos}})\right) p(\mathbf{y}_t^{\text{App}}|\mathbf{x}_t^{(i)}, \mathbf{a}^r)\right] p(\mathbf{x}_t^{(i)}|\mathbf{x}_{t-1}^{(i)})}{q(\mathbf{x}_t^{(i)}|\mathbf{x}_{t-1}^{(i)}, \mathbf{y}_t)}$$

computation of normalized importance weights: $\left\{w_t^{(i)} = \frac{w_t^{(i)}}{\sum_{j=1}^N w_t^{(j)}}\right\}_{i=1}^N$

The results of Figure 2.17 show that the suggested approach allows us to correctly estimate the shape of the mouth over time. In this case, each of the mouths of the learning base is defined by a set $\mathbf{p}_t^d$ of P points of a closed contour around the region from which we extract the color histogram, corresponding to the feature function $\gamma(\mathbf{p}_t^d)$. The state vector is defined by $\mathbf{x}_t = (x_t, y_t, v_t^x, v_t^y, \theta_t, \mathbf{p}_t)$, where (x_t, y_t) is the center of the mouth, (v_t^x, v_t^y) - its speed, θ_t - its orientation and $\mathbf{p}_t$ - the set of control points P. The dynamic model $p(\mathbf{x}_t|\mathbf{x}_{t-1})$ is a second-order autoregressive model. The proposal function is the function that uses fuzzy information, defined in Chapter 3. Every particle $\mathbf{x}_t^{(i)}$, $i = 1, \dots, N$ is, therefore, a possible realization, from which we can determine the membership probabilities defined above and thus compute the weight of a combination of likelihood models. The likelihoods $p(\mathbf{y}_t^{\text{App}}|\mathbf{x}_t^{(i)}, \mathrm{a}^r)$, which we will not detail here, are a combination of color and contour information, reference histograms $\{\mathrm{a}^r\}_{r=1}^R$ that were learned beforehand from the SVM.

Figure 2.17. *Result of the mouth tracking obtained by estimating shapes in the* `Talking face` *sequence*[7]. *The estimated shapes of the mouths are in blue. For a color version of the figure, see www.iste.co.uk/dubuisson/tracking.zip*

7 http://personalpages.manchester.ac.uk/staff/timothy.f.cootes/data/talking_face/talking_face.html.

This approach has proved to be robust for estimating mouth shapes. Note that the formulation that we suggested in equation [2.15] is generalizable to other applications in shape tracking. We could, for instance, consider having a multiple likelihood model to define movement, where the characteristic used would be, for example, a motion vector field.

2.4. Conclusion

We presented in this chapter a certain number of approaches that aim to represent data in an efficient manner, to integrate it into the particle filter with more ease and thus accelerate the computations.

A part of the works presented was concerned with accelerating the computations of the likelihood function when we consider histogram models: some of these methods accelerate its computation significantly [BER 12, DUB 10b, DUB 11a, DUB 12a]. All of these algorithms were more competitive than the IH technique, which is the current reference in the field. An incremental method for computing the distance between histograms, which makes it possible to forgo computing histograms over time (with the exception of those in the first image) [DUB 10a, DUB 11a], was also suggested. This formulation is original from our perspective, as it exploits temporal redundancy, an idea that has not been explored previously. This solution could be further investigated by suggesting other incremental definitions for computing distances or similarity measures between histograms, which can also be used in tracking (diffusion distance or simple histogram intersection). Moreover, extending the suggested approaches in order to have them take into account regions rotated in the image plane will open it up to a broader scope of application. Each of the suggested algorithms allows such an extension, although currently the cost of expensive additional computations (computing intersections, searching in a tree): it is, therefore, necessary to redefine a more general framework, for example, by using different data structures.

The other part of the works was concerned with the representation modes of more complex information, of the sort of object appearance and disappearance [ABE 07b], deformation [ABE 06] or multifeature. We divided the study of this latter type of information by considering the multimodal [ERD 12b], multifragment [ERD 12a] and multiappearance [WID 11a] cases. If this last approach deserves to be examined further in terms of comparative tests and robustness, it provides an original solution to multiappearance tracking, integrated into the particle filter, which aims to separate posture from appearance and to use a likelihood function with multiple models that can be extended to other applications (learning remaining nonetheless necessary). The algorithms dedicated to multimodal and multifragment cases are efficient solutions for tracking a sequence without having to define an architecture in advance (for example, a processing order or confidences set beforehand). These two algorithms adapt to every instant of the contextual conditions and are capable of taking advantage of self-exclusive information to make for tracking of a higher quality than that of the classical methods used until now. However, it could be interesting to use other measurements of quality to update the confidence of the features. For example, for more flexible decision-making, we could use fuzzy measurements in its place, by taking inspiration from works on integrating fuzzy information in the particle filter, which are described in Chapter 3. Finally, rather than using a proposal function by feature, we can consider, in the image of the likelihood function with multiple models suggested in this chapter, to use a proposal function with multiple models, whose weights will be set according to the confidences of the different features. This would allow better exploration of the state space, as this will not restrict diffusion of a particle to using a proposal function defined from a single feature.

The chapter to follow is dedicated to methods for decreasing the quantity of computations in the particle filter by limiting the search area in the state space, that is by focusing on the most interesting areas.

3

Tracking Models That Focus on the State Space

3.1. Introduction

In this chapter, we assume that we have chosen a data representation model. As this representation is defined in a large-sized space, it is impossible to scan it exhaustively. The aim of the approaches presented in this chapter is to suggest algorithms that make it possible to determine which regions of this space should be focused on to limit the complexity of the search and thus refine it.

Note that in this work we are not going to discuss a method class in active expansion on the subjects which we are presently beginning work on: optimization methods, whose aim is specifically to explore the state space to find the distribution modes to estimate, according to a minimizing or maximizing objective function. We can mention among those that were integrated into the particle filter, the Kanade–Lucas–Tomasi (KLT) method [LUC 81] or the *Mean-Shift* [COM 03], both similar in essence. The entire family of approaches also uses heuristics: path relinking [SÁN 05b], scatter search [PAN 08b], genetic algorithms [HAN 05a, AHM 12], evolutionary algorithms and even particle swarm optimization (PSO) [JOH 10, ZHA 10].

However, the problem of this family of approaches is that it often goes beyond the theoretical framework of the particle filter. Indeed, every additional particle diffusion can potentially modify the estimated distribution. Moreover, the meta-heuristics on which they are based do not always have the mathematical properties that provide a rigorous and consistent framework, either in terms of convergence or the accuracy of the results. The Annealed Particle Filter (APF) [DEU 05] is probably the most used and consists of adding iterations (or layers) of simulated annealing to the correction stage. The idea is to divide the correction stage into a set of exploration stages, throughout which we diffuse and subsequently correct particles. This diffusion is more and more concentrated, and the likelihood used for the correction is less and less smoothed. Note that APF makes it possible to remain in the theoretical framework of the particle filter, given the resamplings made after every annealed layer. Moreover, the integrated optimization method has known and proven convergence properties [GRA 06b].

In what follows, we outline various contributions that make it possible to focus in the state space. First, we present two methods for data association, described in section 3.2, which allow to perform multi-object tracking by optimizing the choice of areas to scan. Then, we present in section 3.3 an approach aiming to integrate structural relations, given by fuzzy spatial relations, in the particle filter, which allows more flexible exploration of the space. Finally, we outline in section 3.4 works on conjoint estimation of dynamic and static parameters for $3D$ forms.

3.2. Data association methods for multi-object tracking

Multi-object tracking consists of associating objects with measurements over time in order to obtain a trajectory for each of them [VER 05]. We call measurement, a component of an observation, the information that is extracted from the scene by an external procedure. For example, an observation can contain a set of distances supplied by a sonar sensor. Every distance is hence a measurement and reflects the potential presence of an object. The main difficulty is to determine the correct relations between measurements and objects over time, whose movement models are generally unknown. In our case, data association makes sure that each measurement contained in the observation is associated with at most one object (otherwise, it is associated with the ambient "clutter"). In many cases of complex visual environments, data association is of major importance to the object tracking process and there are two primary families of approach. The deterministic methods (the best known of which is the one using the closest neighboring distance [RON 96]) use a scoring function and resolve many simple association problems [KON 03, SON 05]. The probabilistic methods use *a posteriori* estimation probability and determine the associations by using the probability of error [RAS 01]. The literature is quite abundant on this subject, and we mention here only the best known filters (for more details, we refer the readers to [BAR 88]).

The probabilistic data association filter (PDAF) [COX 93, SIN 74] is an extension of the Kalman filter [BAR 88], where innovation is replaced by a combination of innovations related to every association hypothesis, and weighted by an association probability. It represents an approximation of the Bayesian optimal filter. The latter attributes a probabilistic association of each available measurement to an object of interest [SIN 74, BAR 88]. PDAF is restricted to the case where the initial trajectory position of a single object is known. The measurements are uncertain due to the presence of false alarms. It uses every measurement within the validation region to update the Kalman filter according to the following two hypothesis: (1) a measurement can come from an object or be a false alarm, and (2) at a given instant, an object can be at the origin of no measurements, or one measurement at most.

The joint probabilistic data association filter (JPDAF) [BAR 95, FOR 83] is an extension of the PDAF for multi-object tracking. Its principle is based on the use of a PDAF for each object. The number of objects is known and set beforehand. The only difference between the PDAF and the JPDAF lies in the evaluation of the association probability. JPDAF takes into account every available measurement to update the state of an object. The association vector components are always assumed independent. JPDAF always maintains the approximation hypothesis according to which every *a priori* distribution is a Gaussian distribution. The main downfall of the JPDAF is that it does not allow to realize the initialization and/or the deletion of a track, as it assumes that the number of trajectories is known and set beforehand: changing their number during execution can lead to significant errors. In the basic formulation of the JPDAF, the dynamic and likelihood models are assumed linear and Gaussian; the distribution of the filter is approximated by a Gaussian function. With these hypotheses, updating the state can be obtained via a Kalman filter. For non-linear and non-Gaussian models it was suggested in [SCH 01, SCH 03] to use the Monte Carlo technique with the JPDAF implementation (MC-JPDAF). Hence, a generalization of the latter is proposed in [VER 05a] for the general case of several sensors with arbitrary instrumental distributions.

The multiple hypotheses filter (MHT) [REI 79] is, to this day, probably the most efficient, but also the most expensive in terms of processing time [GEL 04, VER 05a]. Here, the number of objects is assumed unknown and can vary over time. This filter considers all the possible association hypotheses. Making an association hypothesis amounts to stating that a measurement can either come from an object (existing or new) or be a false alarm. The set of association hypotheses is represented by a tree structure where every branch can be characterized by an existence probability. The probability of each hypothesis is updated every time an observation becomes available. Therefore, the algorithmic complexity increases exponentially with the number of received observations. To counter this, MHT eliminates unlikely hypotheses, combines similar hypotheses and keeps the remaining hypotheses. However, rapid elimination of an unlikely hypothesis at a given instant is not always the appropriate action, as this hypothesis can become interesting at a later date.

Finally, the probabilistic multiple hypothesis tracker (PMHT) assumes that an object generates zero or more measurements, contrary to the previously described approaches. Therefore, the estimation can be made only once all of the measures become available, which makes this filter impossible to use in online tracking applications.

Usually, the efficiency of the association depends on the information known beforehand and the available measurements. However, in the general case, this information can be missing, the knowledge of the dynamic can be incomplete, a large time interval can separate two observations, objects can be similar, or deform over time, or a measurement could be equidistant from several objects; there are many problems that make association difficult. None of the association methods listed previously can resolve all of these problems.

In this section we provide a study of two contributions integrated into the particle filter. The first, described in section 3.2.1, is a proposal function that allows to only diffuse particles in the moving regions of the state space and then to attribute them to objects in the scene. In particular, this approach makes it possible to manage the cases of appearance and disappearance, and can work with different modes of acquisition, if the objects are in movement. The second, detailed in section 3.2.2, suggests determining associations between objects and measurements by only considering the past trajectories and the current measurements.

3.2.1. *Particle filter with adaptive classification*

In this approach [DUB 06a, DUB 06b, DUB 09], we define $\mathbf{x}_t = \{\mathbf{x}_t^1, \ldots, \mathbf{x}_t^{M_t}\}$ (here, by M_t we designate the number of objects, as it can change over time), where $\mathbf{x}_t^m = (x_t^m, y_t^m)$ is the state of the object, given by its position in the image. Rather than increasing the size of the sample as a function of the number of objects, we suggest attaching a particle at one instant to a single object, and model the objects with the help of $\{\Omega_t^m\}_{m=1}^{M_t}$ classes, containing the particles which were associated with them.

Hence, at the instant $t-1$, a particle $\mathbf{x}_{t-1}^{(i),m}$ is associated with an object m (or class m) and this object is presented by the ellipse $\mathcal{E}_{t-1}^m$ of which the center $\mathbf{c}_{t-1}^m$ is the weighted average of the particles, or $\mathbf{c}_{t-1}^m = \sum_{i=1}^{N_{t-1}^m} w_{t-1}^{(i),m} \mathbf{x}_{t-1}^{(i),m}$, with N_{t-1}^m the number of the particles associated with the object m ($\sum_{m=1}^{M_{t-1}} N_{t-1}^m = N$), and the directions of the major and the minor axes are defined by the eigenvalues of the variance-covariance matrix of the class. Thus, we define a set of M_{t-1} classes $\{\Omega_{t-1}^m\}_{m=1}^{M_{t-1}}$.

The particles are then propagated with the help of a proposal function combined from several laws: Gaussian functions $\mathcal{N}(\mathbf{x}_{t-1}^m, \Sigma)$ centered on the previous estimated positions of the objects $\mathbf{x}_{t-1}^m$ and the variance-covariance matrix Σ, and a uniform distribution over the notable regions revealed by the Gabor transform over the difference image between the instants $t-1$ and t, or:

$$q(\mathbf{x}_t|\mathbf{x}_{t-1}, \mathbf{y}_t) = \sum_{m=1}^{M_{t-1}} \mathcal{N}(\mathbf{x}_{t-1}^m, \Sigma) + \mathcal{U}(G) \qquad [3.1]$$

where $G = \{\mathcal{I}_t(x,y) : (\sum_\sigma \sum_\alpha |\mathcal{I}_t - \mathcal{I}_{t-1}| \star \mathcal{G}_{\sigma,\alpha}) > S\}$, with $\mathcal{G}_{\sigma,\alpha}$ a Gabor filter defined according to the resolution σ and the orientation α and $\star$ as the convolution operator.

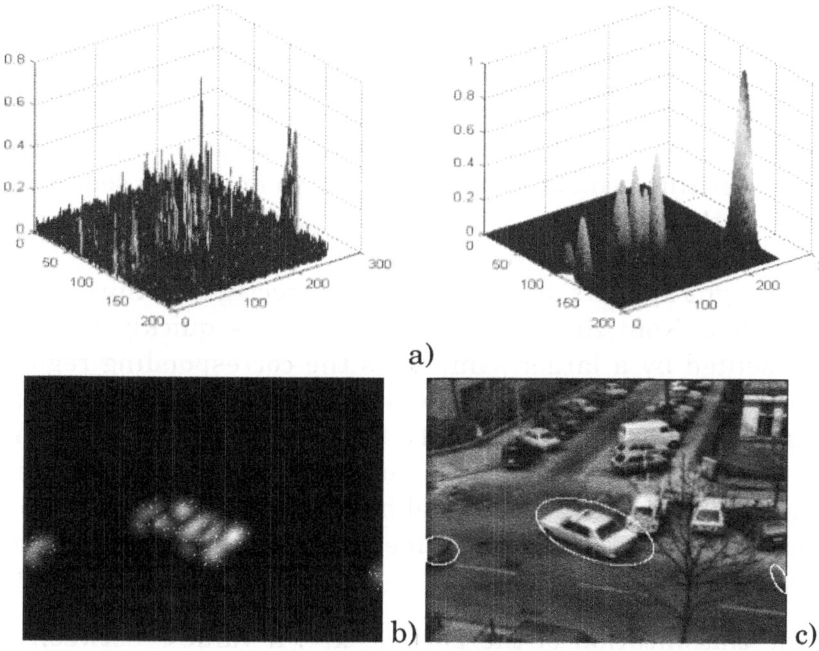

Figure 3.1. *Illustration of the algorithm between the first two images of the sequence* `Taxi`[1]. *a) From left to right, the absolute difference image between two instants and the 3D representation of its response to a Gabor filter bank (along 6 orientations and 2 resolutions, or 12 filters in total). b) The corresponding image of the Gabor response and N = 300 particles positioned uniformly over G. c) The classes $\{\Omega_t^m\}_{m=1}^{M_t}$ obtained, represented by ellipses $\mathcal{E}_t^m$, here M_t = 3. For a color version of the figure, see www.iste.co.uk/dubuisson/tracking.zip*

During the correction stage, every particle is classified, by considering two cases:

1) the particle is in, or relatively close to (as defined), one of the ellipses (classes) $\mathcal{E}_{t-1}^m$. In this case, it is associated at t with

1 http://i21www.ira.uka.de/image_sequences/.

the object m, and the ellipse parameters are updated $\mathcal{E}_t^m$;

2) the particle is too far from all the classes; it is associated with a new class, symbolizing the appearance of a new object or the fact it began moving again.

If, following this classification, a class Ω_{t-1}^m is not updated, then the object associated with it either disappeared or stopped moving. This class is therefore deleted. Figure 3.1 illustrates this principle over the first two images in sequence `Taxi`. In this case, at $t-1$, only the cars on the left and in the center were visible (so there were only two classes). The one on the right appears, and is detected correctly, so a third class is created. Note that an object that moves quickly will be represented by a larger sample, as the corresponding region in G (which characterizes movement) will be spatially larger, thus giving a larger probability for propagating particles in this area. Once a particle is associated with a class, we compute its weight via the likelihood function defined by the Bhattacharyya distance. The method is summarized in the Algorithm 3.1. The function $\texttt{classif}\left(\mathbf{x}_t^{(i)}\right)$ returns the result of the classification of the particle $\mathbf{x}_t^{(i)}$: a value c between 1 and M_{t-1}, or $+\infty$ if the particle does not belong to any class.

Figure 3.2 gives two examples of tracking. In the first example, the number of objects (here, ants) is constant. The particles of the same class are represented by the same color. We can see that, despite the proximity of the ants, association is made correctly (see Figure 3.3 for a comparison of tracking errors obtained with our approach and with the JPDAF filter). The second example shows the capacity of the approach to manage occlusions (class fusions), appearances (creation of classes), disappearances and stops (class deletion), as well as restarting movement (class creation).

Algorithm 3.1: Algorithm of the particle filter with adaptive classification [DUB 09]

Input : set of particles $\{\mathbf{x}_{t-1}^{(i)}, w_{t-1}^{(i)}\}_{i=1}^{N}$, ellipses $\{\mathcal{E}_{t-1}^{m}\}_{m=1}^{M_{t-1}}$, classes $\{\Omega_{t-1}^{m}\}_{m=1}^{M_{t-1}}$

Output : set of particles $\{\mathbf{x}_{t}^{(i)}, w_{t}^{(i)}\}_{i=1}^{N}$, ellipses $\{\mathcal{E}_{t}^{m}\}_{m=1}^{M_{t}}$, classes $\{\Omega_{t}^{m}\}_{m=1}^{M_{t}}$

$\left\{\mathbf{x}_{t-1}^{(i)}, 1/N\right\}_{i=1}^{N} \leftarrow$ multinomial resampling of $\left\{\mathbf{x}_{t-1}^{(i)}, w_{t-1}^{(i)}\right\}_{i=1}^{N}$

$M_t \leftarrow 0$

$\{\Omega_t^m\}_{m=1}^{M_t} \leftarrow \emptyset$

for $i = 1$ **to** N **do**

  Propagation : $\mathbf{x}_t^{(i)} \sim q\left(\mathbf{x}_t|\mathbf{x}_{t-1}^{(i)}, \mathbf{y}_t\right)$ (see equation [3.1])

  $c \leftarrow \texttt{classif}\left(\mathbf{x}_t^{(i)}\right)$

  if $c \leq M_{t-1}$ **then**

    $\Omega_t^c \leftarrow \mathbf{x}_t^{(i)}$

    $\mathbf{x}_t^{c,(i)} \leftarrow \mathbf{x}_t^{(i)}$

  otherwise

    $M_t \leftarrow M_t + 1$

    $\Omega_t^{M_t} \leftarrow \mathbf{x}_t^{(i)}$

    $\mathbf{x}_t^{M_t,(i)} \leftarrow \mathbf{x}_t^{(i)}$

Deleting classes such as $\Omega_t^c = \emptyset$, $c = 1, \ldots, M_{t-1}$

Computation of new parameters of the ellipses $\{\mathcal{E}_t^m\}_{m=1}^{M_t}$ Correction :

$$w_t^{(i)} \propto w_{t-1}^{(i)} \frac{p\left(\mathbf{y}_t|\mathbf{x}_t^{(i)}\right) p\left(\mathbf{x}_t^{(i)}|\mathbf{x}_{t-1}^{(i)}\right)}{q\left(\mathbf{x}_t^{(i)}|\mathbf{x}_{t-1}^{(i)}, \mathbf{y}_t\right)} \qquad i = 1, \ldots, N$$

Normalization : $\left\{w_t^{(i)}\right\}_{i=1}^{N} \leftarrow \left\{\frac{w_t^{(i)}}{\sum_{j=1}^{N} w_t^{(j)}}\right\}_{i=1}^{N}$

3.2.2. *Energetic filter for data association*

The problem of data association is fundamental to temporal multi-object tracking in the domains of surveillance, computer vision and signal processing [BAR 88, COX 93]. The main difficulty is to determine how to associate a measurement with an object. This is made even more difficult if the sensor can generate false alarms, i.e. measurements that do not correspond to any object in the observation region. Moreover, lack of *a priori* knowledge of the number of objects (a number that can change over time), further increases this difficulty.

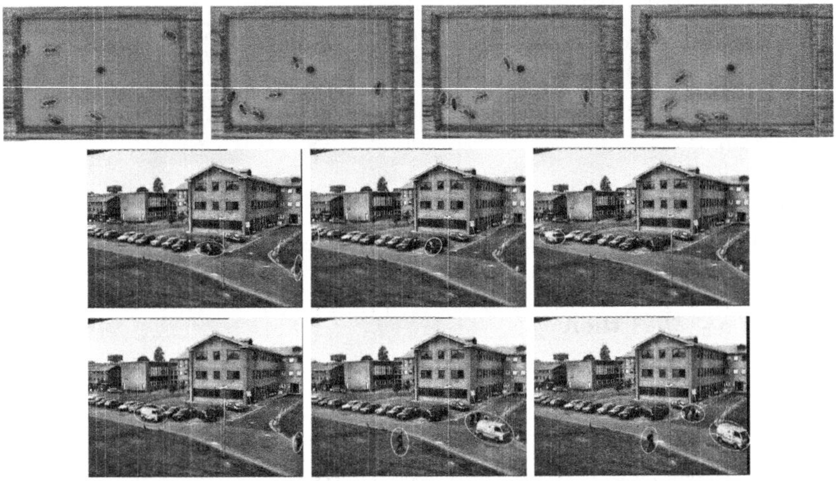

Figure 3.2. *Some illustrations of the results obtained with the approach suggested in [DUB 09]. The top row (sequence* `Ants`[2]*): case of tracking with a constant number of objects, and trajectories that cross. Bottom row (sequence* `Parking`[3]*), several cases, from left to right: a car and a moving pedestrian, disappearance of a pedestrian (on the right), appearance of a new vehicle (on the left) and a vehicle stopping in the middle (its class disappears), appearance of a pedestrian on the right, the vehicle still moving, appearance of a pedestrian in the middle, fusion of the vehicle and the pedestrian classes (on the right) and finally, separation of the vehicle and the pedestrian classes, the single pedestrian continues walking*

A new method for associating measures with objects based on a geometric criterion, called the Energetic Association Filter (EAF), was suggested in [ABE 12]. The idea is to define a term E, depending on the past dynamic of the objects, from which the probability of an association between objects and measurements is deduced. The primary advantage of this approach is that it uses very few parameters, only the last

2 http://www.kinetrack.org/experimental-data.html.

3 http://www.cvg.cs.rdg.ac.uk/PETS2001/pets2001-dataset.html.

three positions of the objects estimated in the image plane, and it successfully manages objects of similar appearance that are equidistant to the same measurement. In what follows, in order to simplify notations, we will assume that the state and observation spaces are the same, and correspond to a Cartesian space (a measurement is therefore a position, just as is the estimated state of the object: this amounts to having an observation function equal to the identity). A more general version of the algorithm is provided in [ABE 12].

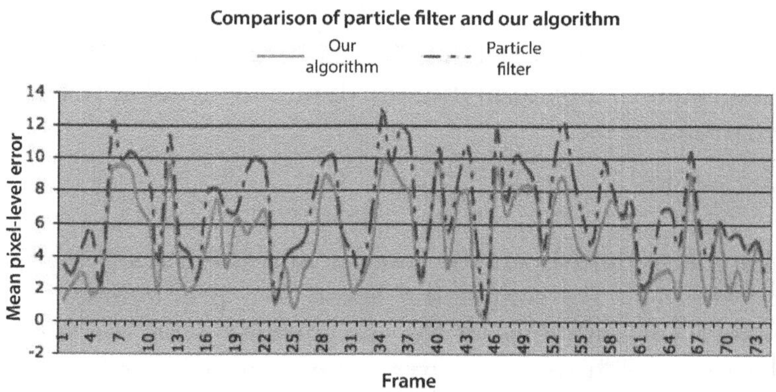

Figure 3.3. *Comparison of tracking errors between the JPDAF filter (in red) an the approach suggested in [ABE 12] (in blue) over the* `Ants` *sequence*

At every instant t, a sensor outputs either zero or an observation y_t, containing at least a measurement of the position y_t^l, $l = 1, \ldots, L$, that can come from an object, or not (in the latter case, it is a false alarm). Every object x_t^m, $m = 1, \ldots, M$ is characterized by its position (x, y) and is associated with a trajectory. We assume then that a measurement can be associated with a trajectory, or not [BAR 88], and that a trajectory contains at most one measure at each instant.

Remember that it is possible to predict a state from its previous estimation with equation [1.1]. We suggest computing the probability at the instant t of an association between the object $\mathbf{x}_t^m$ and the measurement $\mathbf{y}_t^l$ by predicting the position of the object in two different ways at the instant $t+1$; we call these predictions $\left(\mathbf{x}_{t+1}^m\right)^{\#1}$ and $\left(\mathbf{x}_{t+1}^m\right)^{\#2}$, and they are described below as following:

– we do not take into account the current measurement (in other words, we consider that it does not exist). In this case, we predict the state of the object at $t+1$ by the formula:

$$\left(\mathbf{x}_{t+1}^m\right)^{\#1} = f_{t+1}(\mathbf{x}_t^m, \mathbf{u}_{t+1}) \qquad [3.2]$$

– we take into account the current measurement, which we associate with an object. In this case, we predict the state of the object at $t+1$ by applying the dynamic model to the current measurement, or:

$$\left(\mathbf{x}_{t+1}^m\right)^{\#2} = f_{t+1}(\mathbf{y}_t^l, \mathbf{u}_{t+1}) \qquad [3.3]$$

For more details on the models and the parameters used, see [ABE 12]. By comparing these two predictions, we can measure the influence of the measurement on the predicted trajectory. Figure 3.4 gives an example of predicted measurements obtained in two ways: at instant $t+1$, two predictions are made, one without using the current measure $\mathbf{y}_t^l$ (blue trajectory), and the other associating the measurement with its object (red trajectory). During the $t, t+1$ interval, the use of the current measure changes the trajectory significantly. This approach consists of comparing these two trajectories to measure the probability of the object-measurement association.

The score of the association between the object $\mathbf{x}_t^m$ and the measurement $\mathbf{y}_t^l$ (all measurements that did not belong to the validation region of $\mathbf{x}_t^m$ [COX 93] were deleted in advance) is given by the linear combination of the three terms:

$$E(\mathbf{x}_t^m, \mathbf{y}_t^l) = \frac{1}{3}\sum_{i=1}^{3} \alpha_i E^i(\mathbf{x}_t^m, \mathbf{y}_t^l) \qquad [3.4]$$

where $\alpha_i = \left(\sum_{l=1}^{L} E^i(\mathbf{x}_t^m, \mathbf{y}_t^l)\right)^{-1}$, $i = \{1, 2, 3\}$, is a weighting factor.

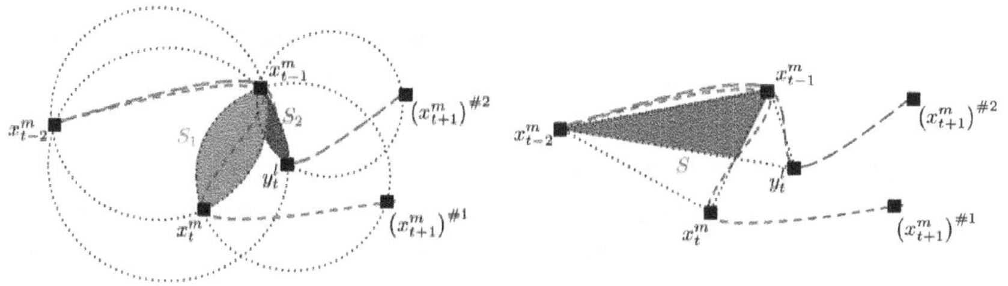

Figure 3.4. *Geometric representation of the trajectory comparison idea. The blue and the red trajectories are obtained respectively with and without the association between the object $\boldsymbol{x}_t^m$ and the current measurement $\boldsymbol{y}_t^l$. On the left, the surfaces S_1 and S_2. On the right, the surface S. For a color version of the figure, see www.iste.co.uk/dubuisson/tracking.zip*

Every term concerns a category of possible movements and their combination allows this approach to associate measurements with objects in a very reliable manner, as we will confirm later. We succinctly describe below their computation from geometric criteria only (areas and distances). The details are given in [ABE 12]:

– the distance term E^1 (case of linear translation movement) corresponds simply to the Mahalanobis distance between the measurement and the object at the instant t, given by $E^1(\mathbf{x}_t^m, \mathbf{y}_t^l) = \sqrt{\left(\mathbf{y}_t^l - \mathbf{x}_t^m\right)^T (\Sigma_t^m)^{-1} \left(\mathbf{y}_t^l - \mathbf{x}_t^m\right)}$, where Σ_t^m is a variance-covariance matrix. This distance generates association errors if the measurements and the predictions are equidistant or if the trajectories cross, which requires the introduction of a second term;

– the evolution term E^2 is adapted to the case of complex dynamics (oscillatory movement or non-constant speed) and measures the proximity between two possible trajectories (i.e. with and without using the current measurement). It is given by $E^2(\mathbf{x}_t^m, \mathbf{y}_t^l) = S_1 - S_2$, where S_1 is the intersection of the two circles circumscribing the triangles $\left(\mathbf{x}_{t-2}^m, \mathbf{x}_{t-1}^m, \mathbf{x}_t^m\right)$

and $\left(\mathrm{x}_{t-1}^{m}, \mathrm{x}_{t}^{m}, (\mathrm{x}_{t+1}^{m})^{\#1}\right)$ and S_2 is the intersection of the two circles circumscribing the triangles $\left(\mathrm{x}_{t-2}^{m}, \mathrm{x}_{t-1}^{m}, \mathrm{y}_{t}^{l}\right)$ and $\left(\mathrm{x}_{t-1}^{m}, \mathrm{y}_{t}^{l}, (\mathrm{x}_{t+1}^{m})^{\#2}\right)$ (see Figure 3.4).

The similarity between the two dynamic models is highest when $E^2(\mathrm{x}_t^m, \mathrm{y}_t^l)$ is lowest (i.e. the surfaces S_1 and S_2 are equal). $E^2(\mathrm{x}_t^m, \mathrm{y}_t^l)$ decreases when the object x_t^m and the measurement y_t^l are close, as are the predictions $(\mathrm{x}_{t+1}^{m})^{\#1}$ and $(\mathrm{x}_{t+1}^{m})^{\#2}$, as well as their associated dynamic model, proving that y_t^l comes from the object x_t^m. However, if x_t^m and y_t^l are diametrically opposed, or even opposed with respect to the $\left(\mathrm{x}_{t-2}^{m}, \mathrm{x}_{t-1}^{m}\right)$ axis, then $S_1 \approx S_2$ and $E^2(\mathrm{x}_t^m, \mathrm{y}_t^l) \approx 0$. To manage this particular case, which corresponds to the crossing of two trajectories, we introduce a third term:

– the temporal proximity term E^3 measures the proximity of two trajectories over time and is given by $E^3(\mathrm{x}_t^m, \mathrm{y}_t^l) = S^{-1}$, where S is the common area between the triangles $\left(\mathrm{x}_{t-2}^{m}, \mathrm{x}_{t-1}^{m}, \mathrm{y}_{t}^{l}\right)$ and $\left(\mathrm{x}_{t-2}^{m}, \mathrm{x}_{t-1}^{m}, \mathrm{x}_{t}^{m}\right)$ (see Figure 3.4, pink area). The closer x_t^m and y_t^l, the more S increases: measuring this term at instant t makes it possible to add information about previously similar trajectories to the final score E.

The global normalized scores are then computed for each measurement, with equation [3.4], which gives an association score between the object x_t^m and each measurement y_t^l, which we call $\beta^{ml} = 1 - E(\mathrm{x}_t^m, \mathrm{y}_t^l)$. The measurement that has the greatest score will be the one with the object. We give the general workings of the approach suggested in Algorithm 3.2. This approach proved itself to be very efficient in associating objects with measurements in the case of varied movement, even if the objects are similar. It was introduced into the particle filter as well, which resulted in the Energetic Particle Filter (EPF) [ABE 07a]. In the particle filter algorithm, the association stage naturally comes into play just after the stage that predicts the new positions of the objects. The correction is therefore made after associating measurements with the objects.

Algorithm 3.2: Algorithm of the Energetic Association Filter (EAF) [ABE 07a]

Input : previously estimated states of the objects $\{\mathbf{x}_{t-2}^m, \mathbf{x}_{t-1}^m, \mathbf{x}_t^m\}_{m=1}^M$, observation $\mathbf{y}_t = \{y_t^l\}_{l=1}^L$.
Output : association between objects and measurements
for $m = 1$ **to** M **do**
Compute the prediction without association:
$$(\mathbf{x}_{t+1}^m)^{\#1} = f_{t+1}(\mathbf{x}_t^m, \mathbf{u}_{t+1})$$
for $l = 1$ *at L, valid* $\mathbf{y}_t^l$ *[COX 93]* **do**
Compute the prediction with association:
$$(\mathbf{x}_{t+1}^m)^{\#2} = f_{t+1}(\mathbf{y}_t^l, \mathbf{u}_{t+1})$$
Compute distance term
$$E^1(\mathbf{x}_t^m, \mathbf{y}_t^l) = \sqrt{(\mathbf{y}_t^l - \mathbf{x}_t^m)^T (\Sigma_t^m)^{-1} (\mathbf{y}_t^l - \mathbf{x}_t^m)}$$
Compute evolution term $E^2(\mathbf{x}_t^m, \mathbf{y}_t^l) = |S_1 - S_2|$
Compute temporal proximity term $E^3(\mathbf{x}_t^m, \mathbf{y}_t^l) = \frac{1}{S}$
for $l = 1$ **at** L, *valid* $\mathbf{y}_t^l$ **do**
Compute normalized global scores:
$$E(\mathbf{x}_t^m, \mathbf{y}_t^l) = \frac{1}{3}\sum_{i=1}^{3} \alpha_i E^i(\mathbf{x}_t^m, \mathbf{y}_t^l)$$
$$\text{with} \quad \alpha_i = \frac{1}{\sum_{l=1}^L E^i(\mathbf{x}_t^m, \mathbf{y}_t^l)}$$
Compute a posteriori probabilities of association between the measurements $\mathbf{y}_t^l$ and the object $\mathbf{x}_t^m$: $\beta^{ml} = 1 - E(\mathbf{x}_t^m, \mathbf{y}_t^l)$
$\beta^{ml^\bullet} = \max_l \beta^{ml}$
$\mathbf{y}_t^{l^\bullet}$ is associated to the object $\mathbf{x}_t^m$

Figure 3.5 gives an example of tracking six objects (ants). We can see that, even when the ants are close and cross each other, tracking remains correct, as shown by the estimated trajectories in the bottom image. The comparison of the trajectories estimated by the Standard Particle Filter (SPF), the JPDAF and our EPF is given in Figure 3.6: we can see that EPF has better stability in relation to the other approaches. In this sequence, the average tracking error for all the ants (average of the distances between the estimated position for each ant and their actual position in the image) given by this approach is 4.3 pixels, against 7.5 pixels for the JPDAF and 14.2 pixels for the SPF.

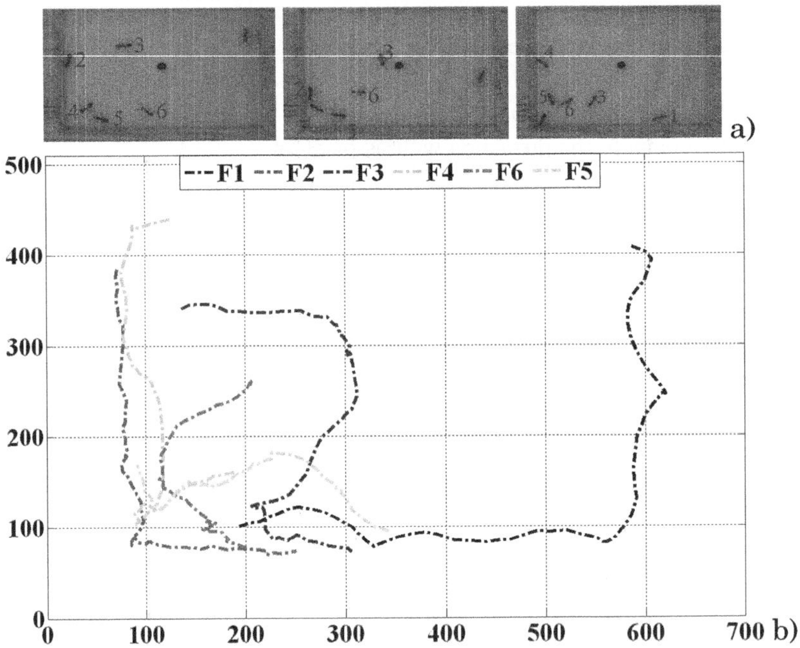

Figure 3.5. *Multi-object tracking int sequences* `Ants`[4]. *a) Images 10, 25 and 45. b) estimated trajectories for the six ants throughout the entire sequence, with $N = 500$. For a color version of the figure, see www.iste.co.uk / dubuisson / tracking.zip*

We have just presented two approaches dedicated to multi-object tracking, the first maintains a classification of the particles over time, and the second computes an association probability at the current instant as a function of the past trajectory of the objects. In the following section, we interest ourselves in the introduction of fuzzy spatial information into the particle filter to make the tracking more flexible, and thus improve it.

4 http://www.kinetrack.org/experimental-data.html.

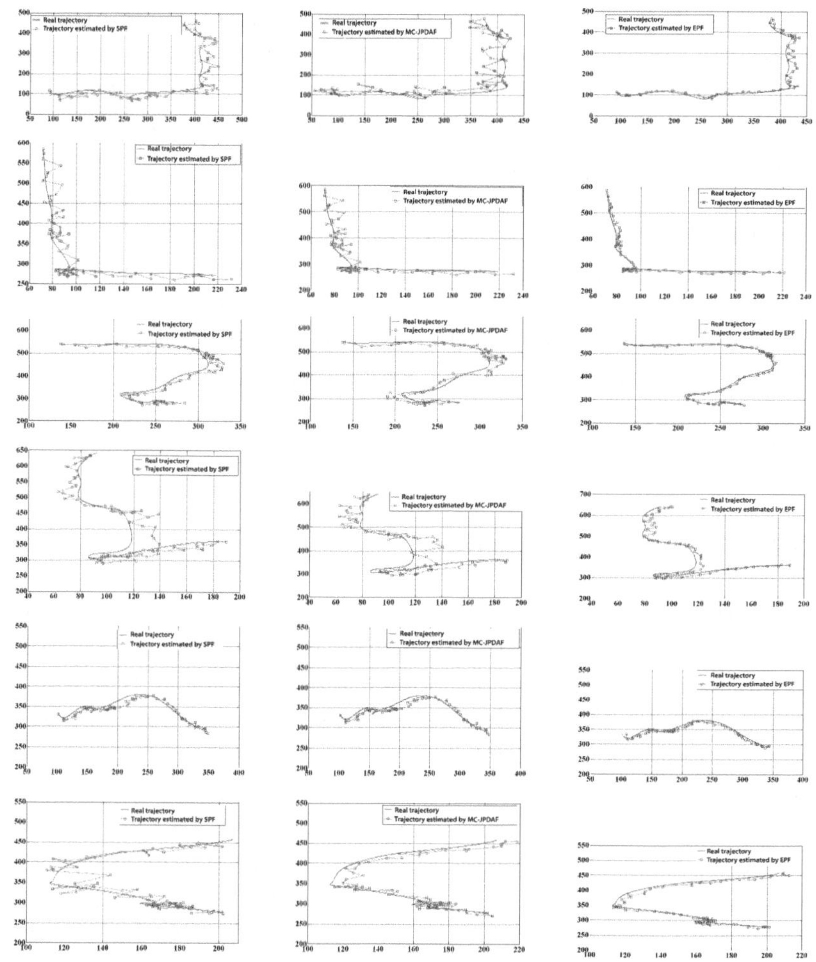

Figure 3.6. *Trajectories estimated for each ant (top to bottom, ants 1 to 6 of Figure 3.5.(a)), obtained, from left to right, by an SPF, the JPDAF and our EPF, for N =500. For a color version of the figure, see www.iste.co.uk/dubuisson/tracking.zip*

3.3. Introducing fuzzy information into the particle filter

Fuzzy spatial information makes it possible to represent the spatial configuration of an object or of several objects in the scene. The idea is to model fuzzy spatial information, to define the fuzzy spatio-temporal relations and to integrate them to particle filter tracking. Therefore the aim is to introduce imprecise structural information, either between objects or between the positions of an object at different instants. These vague spatial relations are modeled by fuzzy sets. In a few words, we first present the principle behind fuzzy modeling, then fuzzy spatial relations, respectively in sections 3.3.1 and 3.3.2. Then, in section 3.3.3, we explain how they can be integrated into the probabilistic scheme of the particle filter. Finally, we give three applied examples in tracking objects (sections 3.3.3.1 and 3.3.3.2) or the shape of an object (section 3.3.3.3).

3.3.1. *Fuzzy representation*

The fuzzy set theory makes it possible to model qualitative knowledge, vague or imprecise, often expressed linguistically [DUB 80]. In object tracking, it is possible to model certain object characteristics or behaviors through fuzzy representation. In this case, the modeled information is deterministic, but vague, making fuzzy sets appropriate for its representation, as they inherently allow to express this imprecision. Thus, it is possible to describe the concept of object volume not by an exact value, but a fuzzy set with a well-chosen granularity, of the *Small*, *Medium* or *Large* type, which has an impact on the precision of the representation. Likewise, it is possible to explore the *fuzzy concept* of proximity between two objects, characterizing terms as *Close* or *Far*. The semantics of these terms is therefore defined by fuzzy sets.

DEFINITION 3.1 (Fuzzy information).– A fuzzy concept is defined by a family of K fuzzy sets $\{\mathcal{A}_k\}_{k=1}^{K}$, defined by membership functions $\{\mu_k\}_{k=1}^{K}$, such that $\forall k \in \{1, \ldots, K\}, \mu_k : \mathcal{O} \rightarrow [0, 1]$, $\mathcal{O} \subset \mathbb{R}^{n_o}$, with space $\mathcal{O}$ of size n_o, over which the concept is defined. The function family

$\{\mu_k\}_{k=1}^K$ corresponds to a linguistic variable in which every function μ_k gives the semantics of a possible linguistic value of a variable. Every set corresponds to a possible value of a fuzzy concept, and contains elements that are possible representations of this value. To each element $o \in \mathcal{O}$ is associated a degree of membership $\mu_k(o)$ with a fuzzy set $\mathcal{A}_k$.

Let us take an example where we consider a fuzzy concept of orientation. The $K = 4$ fuzzy sets could be $\mathcal{A}_1 = North$, $\mathcal{A}_2 = East$, $\mathcal{A}_3 = South$ and $\mathcal{A}_4 = West$ ($\mathcal{O} = 2\pi$), each defined by a membership function $\{\mu_k\}_{k=1}^4$ (illustrated by a later example). In this case, the angle $\pi/2$ has a high degree of membership to the fuzzy set $\mathcal{A}_1 = North$ ($\mu_1(\pi/2)$ has a high value). In contrast, it reflects a strong improbability of the linguistic value $\mathcal{A}_3 = South$ ($\mu_3(\pi/2)$ has a low value).

We do not detail how to model membership functions associated with a certain fuzzy concept here; instead we direct the reader to [DUB 80], but remark that they can be fixed, either by *a priori* knowledge or by learning. These fuzzy sets can then interact by using strategies for fusing fuzzy functions [BLO 96]. Figure 3.7 gives the representation of a spatial entity by a fuzzy set in the image domain.

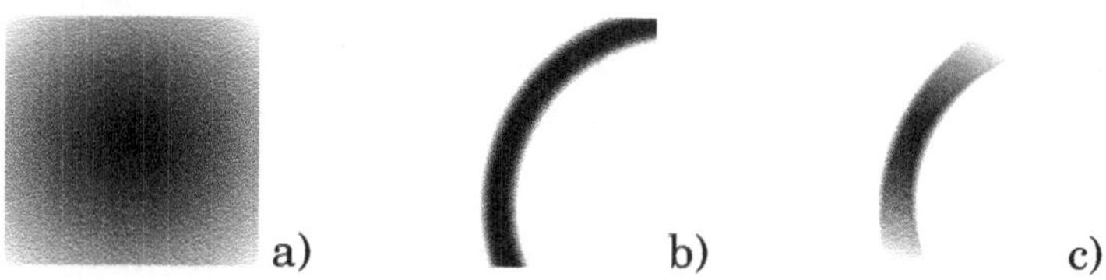

Figure 3.7. *Representation of a fuzzy set in the image domain (derived from [WID 10a]). a) A fuzzy ball, b) a constraint (here, spatial localization) and, c) with the help of a t-norm (fuzzy conjunction), the constraint result (restriction of the fuzzy object represented in (a) at the part that satisfies the spatial localization constraint represented in (b)). The grayscale level associated with each pixel reflects the degree of membership of the element that it represents to a fuzzy set, a darker gray indicating a stronger belonging*

3.3.2. *Fuzzy spatial relations*

The spatial relations supply important structural information on or between objects evolving in the video sequence. Thus, we could for example define relations on distances (*Close*, *At medium distance*, *Far*), directions (*West*, *South*) or even movement (*Up*, *Left*). A spatial relation is considered as a linguistic variable, defined by the functions $\{\mu_k\}_{k=1}^{K}$, that can take a certain number of values (that is, belonging to a function μ_k) [ZAD 75].

Let us come back to the example of the fuzzy orientation concept, for which we can consider eight linguistic values: *East*, *North-East*, *North*, *North-West*, *West*, *South-West*, *South*, *South-East*. Every value can be defined according to semantics defined by a membership function that is trapezoidal on the angle axis, and the set of membership functions forms a fuzzy partition over $[0, 2\pi]$. Some of these membership functions can be superimposed, for instance *North* and *North-East*. Figure 3.8 gives an illustration of this uniform fuzzy partition defined in the parameter space.

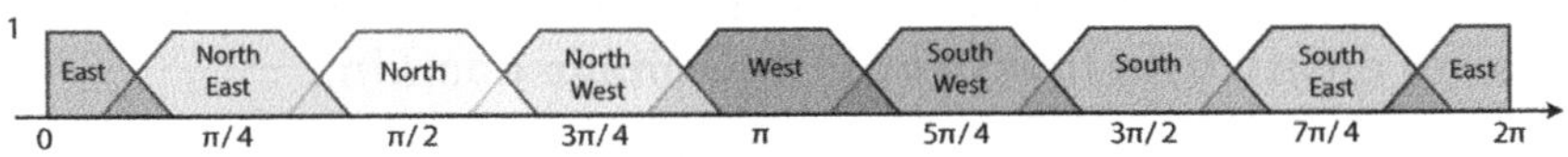

Figure 3.8. *Example of a fuzzy partition describing an angular spatial relation, with $K = 8$ fuzzy sets (extracted from [WID 10a])*

These relations can also be represented in the spatial domain [BLO 05], as illustrated in Figure 3.9, by using different types of fuzzy functions. Note that the representation of fuzzy sets in the spatial domain (of the image) is only given here for illustrative purposes, as in practice, we only manipulate them in the variable domain ($[0, 2\pi]$, $\mathbb{R}^+$, ...).

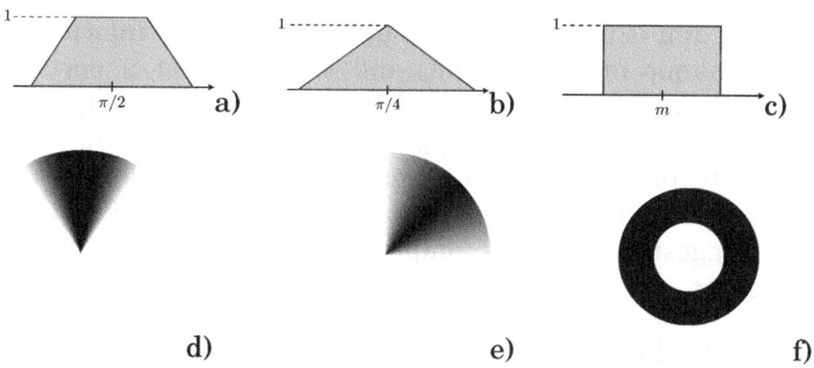

Figure 3.9. *Fuzzy spatial relations represented in the variable (a–c) and image (d–f) domain. A dark value represents maximal belonging; (a,d) North value of the fuzzy concept of orientation by using a trapezoidal shape, (b,e) North-East value of the fuzzy concept of orientation by using a triangular shape and (c,f) Medium distance value of the fuzzy concept of orientation by using a binary shape (extract from [WID 10a])*

3.3.3. *Integration of fuzzy spatial relations into the particle filter*

Until now, the introduction of fuzzy information into the particle filter has only been touched upon very lightly. In general, this essentially concerns the fuzzy fusion of information [FAU 06, GAR 03, KIM 07], which is not necessarily related to the methodological context of tracking, but rather a way of managing information on input or output. In [CAI 08], the number of particles is automatically adapted by using fuzzy logic rules. In [WU 08a], the fuzzy functions are approximated by a particle filter algorithm. The dynamic model, as well as its parameters, are selected in [KAM 05] with fuzzy logics.

In contrast to the previously listed approaches, the fuzzy information integrated here is of spatial nature [WID 10a, WID 11b] and makes it possible to describe the evolution of an object over time. Formally, it is integrated by the continuous/discrete hybrid model as dynamic unknowns of the system, which constitutes an originality. This means that its aim is to describe, through fuzzy spatial information, the dynamic of the system, and thus suggest a pertinent simulation of particles throughout the prediction stage. This also distinguishes this approach from the others which integrate fuzzy spatial information during the updating of the weights, via likelihood. Finally, an importance function is also suggested, which is adapted to the chosen type of fuzzy modeling.

To introduce spatial relations into the particle filter, in the state vector $\mathbf{x}_t$, we need to separate the positional and the dynamic variables, as the latter will be represented by fuzzy concepts. We hence introduce a vector $\mathbf{r}_t = \{\mathbf{r}_t^r\}_{r=1}^R$, whose every component represents a parameter of the dynamic of the system (orientation, velocity, etc.). We will then assume that the components are independent from each other. If we assume that they are related to a dynamic of a continuous random walk, the equation system 1.1 of the filter can be re-written as follows:

$$\begin{cases} \mathbf{x}_t = f_t(\mathbf{x}_{t-1}, \mathbf{r}_{t-1}, \mathbf{u}_t) \\ \mathbf{r}_t^r = \mathbf{r}_{t-1}^r + \mathbf{w}_t^r \quad \forall r \in \{1, \ldots, R\} \\ \mathbf{y}_t = g_t(\mathbf{x}_t, \mathbf{v}_t) \end{cases} \qquad [3.5]$$

with $\{\mathbf{w}_t^r\}_{r=1}^R$ as independent white noise.

However, in order to introduce fuzzy spatial information into the dynamic of the system, we will rather assume that every parameter $\mathbf{r}_t^r$ is associated with a fuzzy concept defined by fuzzy sets $\{\mathcal{A}_k^r\}_{k=1}^{K^r}$, themselves corresponding to fuzzy functions $\{\mu_k^r\}_{k=1}^{K^r}$. For example, if $\mathbf{r}_t^1$ is a vector containing a velocity parameter $\mathbf{r}_t^1 = (v_t^x, v_t^y)$, then we can consider eight linguistic vectors $\mathcal{A}_k^1$ for the couple (v_t^x, v_t^y), from *North-West* to *South-East*, modeling the relative position of the object at

t, given the one it had at $t-1$. We can now modify equation [3.5] in the following way:

$$\begin{cases} \mathbf{x}_t = f_t(\mathbf{x}_{t-1}, \mathbf{r}_t, \mathbf{c}_t, \mathbf{u}_t) \\ \mathbf{y}_t = g_t(\mathbf{x}_t, \mathbf{c}_t, \mathbf{v}_t) \end{cases} \qquad [3.6]$$

where $\mathbf{c}_t = \{\mathbf{c}_t^r\}_{r=1}^R$ is a vector indicating the indices of fuzzy sets $\{\mathcal{A}_k^r\}_{k=1}^{K^r}$ to which the components $\mathbf{r}_t^r$ of the vector of parameters $\mathbf{r}_t$ are most likely to belong. For example, if $\mathbf{r}_t^1 = (v_t^x, v_t^y)$, then $\mathbf{c}_t^1 \in \{$*North-West*, *North*, *North-East*, *West*, *East*, *South-West*, *South*, *South-East*$\}$.

In equation [3.6], we no longer assume that the evolution of $\mathbf{r}_t$ is a random step depending on $\mathbf{r}_{t-1}$, but rather $\mathbf{r}_t$ is related to the fuzzy set indicated by $\mathbf{c}_t$, which restricts the domain of its definition (via its membership function), and that $\mathbf{r}_t$ is distributed uniformly over this set. The evolution of vector $\mathbf{c}_t$ is, for its part, defined with a transition matrix defined from fuzzy sets and can be seen as a hopping pattern. Note that, contrary to equation [3.5], in equation [3.6], we use $\mathbf{r}_t$ and not $\mathbf{r}_{t-1}$ in order to estimate $\mathbf{x}_t$. This makes it possible to increase the reliability of the approach, but in return requires estimating $\mathbf{c}_t$ and $\mathbf{r}_t$ before $\mathbf{x}_t$.

For each parameter $\mathbf{r}_t^r$, we define a fuzzy set of possible events $\mathcal{Z}^r$ by using a fuzzy set aggregation formula $\{\mathcal{A}_k^r\}_{k=1}^{K^r}$, therefore defining a membership function $\mu_{\mathcal{Z}}^r$. The latter makes it possible to retain as a set of definitions of $\mathbf{r}_t^r$ the set $\{\mathbf{r}_t^r : \mu_{\mathcal{Z}}^r(\mathbf{r}_t^r) > 0\}$, in other words, only the values of $\mathbf{r}_t^r$ that do not turn out to be totally improbable. We can then define a set of possible events for $\mathbf{r}_t$: $\mathcal{Z} = \{\mathcal{Z}^1, \ldots, \mathcal{Z}^R\}$ defined by the fuzzy membership functions $\{\mu_{\mathcal{Z}}^1, \ldots, \mu_{\mathcal{Z}}^R\}$.

Finally, to get back to the probabilistic framework of the particle filter, we consider a fuzzy set as a fuzzy event $\mathcal{A}$, whose the probability is given by [ZAD 68]:

$$\mathbb{P}(\mathcal{A}) = \int_{\mathbb{R}^n} \mu_{\mathcal{A}}(\mathbf{x}) d\mathbb{P} = \mathbb{E}[\mu_{\mathcal{A}}] \qquad [3.7]$$

where $\mu_{\mathcal{A}} : \mathbb{R}^n \to [0,1]$ is the membership function according to which the event is defined. It is possible, in a similar way, to define the joint distribution of two fuzzy events (by using the t-norm product as a fuzzy conjunction operator) or a conditional distribution [ZAD 68]. Given that we consider the components of $\mathbf{r}_t$ as independent, we can also consider those from $\mathbf{c}_t$ as independent. Thus, we can deduce the *a priori* transition distribution of $\mathbf{c}_t$:

$$\mathbb{P}(\mathbf{c}_t = (k^1, \ldots, k^R)|\mathbf{c}_{t-1} = (h^1, \ldots, h^R), \mathcal{Z}) \\ = \prod_{r=1}^{R} \mathbb{P}(\mathbf{c}_t^r = k^r|\mathbf{c}_{t-1}^r = h^r, \mathcal{Z}^r) \quad [3.8]$$

with $\mathbb{P}(\mathbf{c}_t^r = k^r|\mathbf{c}_{t-1}^r = h^r, \mathcal{Z}^r)$ the transition matrix for the rth parameter. We can show that, under certain hypotheses, this distribution is defined by the degree of intersection between the fuzzy sets [WID 10a]. Assuming the parameter vector $\mathbf{r}_t$ conditionally independent from $\mathbf{r}_{t-1}$ given $\mathbf{c}_t$, we get:

$$\mathbb{P}(\mathbf{r}_t \in d\mathbf{r}_t|\mathbf{c}_t = (k^1, \ldots, k^R), \mathbf{r}_{t-1}, \mathcal{Z}) \\ = \prod_{r=1}^{R} \mathbb{P}(\mathbf{r}_t^r \in d\mathbf{r}_t^r|\mathbf{c}_t^r = k^r, \mathcal{Z}^r) \quad [3.9]$$

with $d\mathbf{r}_t$ an infinitesimal fuzzy support event ($\mu_{d\mathbf{r}_t}(\mathbf{r}) = 1$ if $\mathbf{r} \in [\mathbf{r} - \frac{d\mathbf{r}_t}{2}, \mathbf{r} + \frac{d\mathbf{r}_t}{2}]$, 0 otherwise). $\mathbb{P}(d\mathbf{r}_t^r|\mathbf{c}_t^r = k^r, \mathcal{Z}^r)$ is the conditional distribution of $\mathbf{r}_t^r$ knowing $\mathbf{c}_t^r$ and $\mathcal{Z}^r$, proportional to the ratio between the degree of membership of $\mathbf{r}_t^r$ in the kth fuzzy set $\mathcal{A}_k^r$ and its degree of membership in the fuzzy set of possible events $\mathcal{Z}^r$, whose probability is defined in equation [3.7] (see [WID 11b] for more details). This probability is precisely what we need to define our particle filter.

The final graphical model is illustrated in Figure 3.10. The distribution of the *a posteriori* trajectory filter, conditioned by the set of possible events $\mathcal{Z}$, can be approximated by the Algorithm 3.3, where we set $\mathbf{s}_t^{(n)} = \left(\mathbf{x}_t^{(n)}, \mathbf{r}_t^{(n)}, \mathbf{c}_t^{(n)}\right)^T$ in order to simplify the notations.

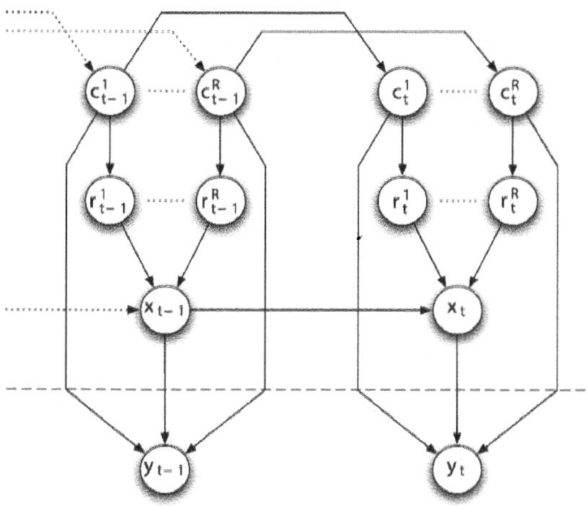

Figure 3.10. *Graphical model of the suggested approach (extracted from [WID 10a]). The random vectors located above the dotted line are hidden and only the observations are visible. Given $\mathbf{c}_t$, the parameter vector $\mathbf{r}_t$ is conditionally independent from $\mathbf{r}_{t-1}$*

The proposal function integrates the last observation, as well as the indices of $a + 1$ (a being a parameter to set) last fuzzy sets $\mathbf{c}_{t-1-a:t-1}$ estimated for the object trajectory, which makes it possible to estimate the current indices $\mathbf{c}_t$ with more precision. Hence, the suggested importance function features adaptive behavior that makes it possible to manage sudden changes in the dynamic, such as outlying values, while using the trajectory information to overlook local ambiguities. For example, by using the trajectory of the indices of the fuzzy functions and considering a constant velocity model, in the presence of erratic movement, the importance function tends to generate uniform exploration of the parameter state space, while in the case of a very linear sequence, a specific region is more likely to be considered. This remains conditioned by the last observation, and therefore by the likelihoods contained in the different parameter hypotheses. The simulation process for a given fuzzy set, as well as details on modeling the proposal and transition functions are given in [WID 10a]. In the section that follows, we give some applied examples.

Algorithm 3.3: Particle filter with integrated fuzzy spatial information [WID 10a, WID 11b]

Input : set of particles $\{\mathbf{s}_{0:t-1}^{(i)}, w_{t-1}^{(i)}\}_{i=1}^{N}$

Output : set of particles $\{\mathbf{s}_{0:t}^{(i)}, w_t^{(i)}\}_{i=1}^{N}$

$\left\{\mathbf{s}_{0:t-1}^{(i)}, 1/N\right\}_{i=1}^{N} \leftarrow$ multinomial resampling of $\left\{\mathbf{s}_{0:t-1}^{(i)}, w_{t-1}^{(i)}\right\}_{i=1}^{N}$

for $i = 1$ **to** N **do**

Propagation:

$$\mathbf{c}_t^{(i)} \sim q(\mathbf{c}_t|\mathbf{c}_{t-1-a:t-1}^{(i)}, \mathbf{r}_{t-1}^{(i)}, \mathbf{x}_{t-1}^{(i)}, \mathbf{y}_t, \mathcal{Z})$$

$$\mathbf{r}_t^{(i)} \sim \mathbb{P}(\mathbf{r}_t \in d\mathbf{r}_t|\tilde{\mathbf{c}}_t^{(i)}, \mathcal{Z})$$

$$\mathbf{x}_t^{(i)} \sim q(\mathbf{x}_t|\mathbf{x}_{t-1}^{(i)}, \tilde{\mathbf{r}}_t^{(i)})$$

Set $\mathbf{s}_t^{(i)} = (\mathbf{x}_t^{(i)}, \mathbf{r}_t^{(i)}, \mathbf{c}_t^{(i)})$ and $\mathbf{s}_{0:t}^{(i)} = (\mathbf{s}_{0:t-1}^{(i)}, \mathbf{s}_t^{(i)})$

Correction:

$$w_t^{(i)} = w_{t-1}^{(i)} \frac{p(\mathbf{y}_t|\mathbf{x}_t^{(i)}, \mathbf{r}_t^{(i)}, \mathbf{c}_t^{(i)}) p(\mathbf{x}_t^{(i)}|\mathbf{x}_{t-1}^{(i)}, \mathbf{r}_t^{(i)}) \mathbb{P}(\mathbf{c}_t^{(i)}|\mathbf{c}_{t-1}^{(i)}, \mathcal{Z})}{q(\mathbf{x}_t^{(i)}|\mathbf{x}_{t-1}^{(i)}, \mathbf{r}_t^{(i)}) q(\mathbf{c}_t^{(i)}|\mathbf{c}_{t-1-a:t-1}^{(i)}, \mathbf{r}_{t-1}^{(i)}, \mathbf{x}_{t-1}^{(i)}, \mathbf{y}_t, \mathcal{Z})}$$

Normalization: $\left\{w_t^{(i)}\right\}_{i=1}^{N} \leftarrow \left\{\frac{w_t^{(i)}}{\sum_{j=1}^{N} w_t^{(j)}}\right\}_{i=1}^{N}$

3.3.3.1. *Application to tracking an object with erratic movements*

Algorithm 3.3 was applied in the context of tracking an object in synthetic sequences (in which several trajectories were generated). We present here the results obtained in the real sequence `Helicopter`, featuring a remote-controlled helicopter subject to very erratic movements and whose aspect changes over time (change in scale and appearance).

The elliptic region around the object is estimated, that is, the unknown states of the positions and the dynamics $\mathbf{x}_t = (x_t, y_t)^T$ and $\mathbf{r}_t = (\dot{x}_t, \dot{y}_t, \theta_t, w_t, h_t)^T$, with θ_t the object orientation, w_t the width and h_t the height of the ellipse. The dynamic model is a constant speed, and the velocity, orientation and dimension parameters are estimated by our fuzzy formalism. The likelihood is modeled as an independent combination of a regional and a contour component, invariant to changes in scale and rotation. It is detailed in [WID 10a, WID 11b].

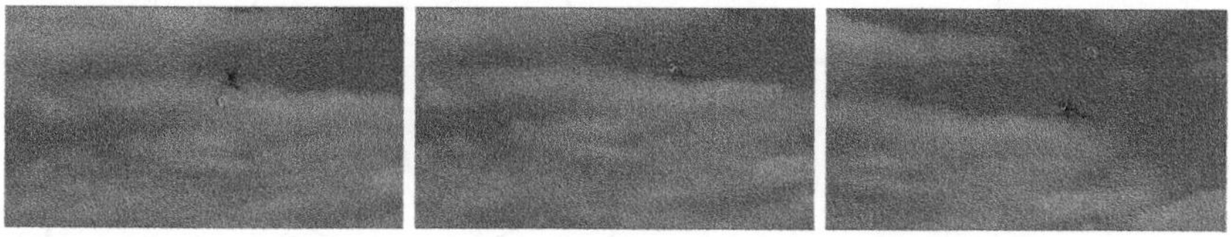

Figure 3.11. *`Helicopter` sequence*[5]. *Close-ups on the images with ellipses estimated by different filters for $N = 50$ particles at instants 12, 18 and 78. Green: SIR, red: AUX-PF, violet: MMPF, yellow: AUX-MMPF, brown: FS-PF (model in [WID 11b] without importance function) and blue: PFS-PF (model in [WID 11b] with importance function). For a color version of the figure, see www.iste.co.uk / dubuisson / tracking.zip*

The model was compared to an SIR filter, an auxiliary particle filter (AUX-PF), a multi-model particle filter (MMPF) and an auxiliary multi-model particle filter (AUX-MMPF). Figure 3.11 shows several tracking results, where we can see that this approach, using an importance function defined by fuzzy functions, always follows the object. The trajectories estimated by each method, as well as the ground truth, are given in Figures 3.12(a) and 3.12(b). We can see in Figure 3.12(c) that the approach remains very competitive in comparison to the other ones tested, as it decreases both the processing time and the estimation error up to $N = 300$. Beyond this, we find that it still reduces the estimation error, however the processing time gains are no longer systematic if we compare with, for example, the MMPF or SIR.

5 http://www.youtube.com/watch?v=PQf404RumPE.

3.3.3.2. *Application to multi-object tracking*

Without giving the details of the modeling, which can be found in [WID 10a, WID 11b], we show the interest of introducing fuzzy spatial relations into the particle filter for multi-object tracking.

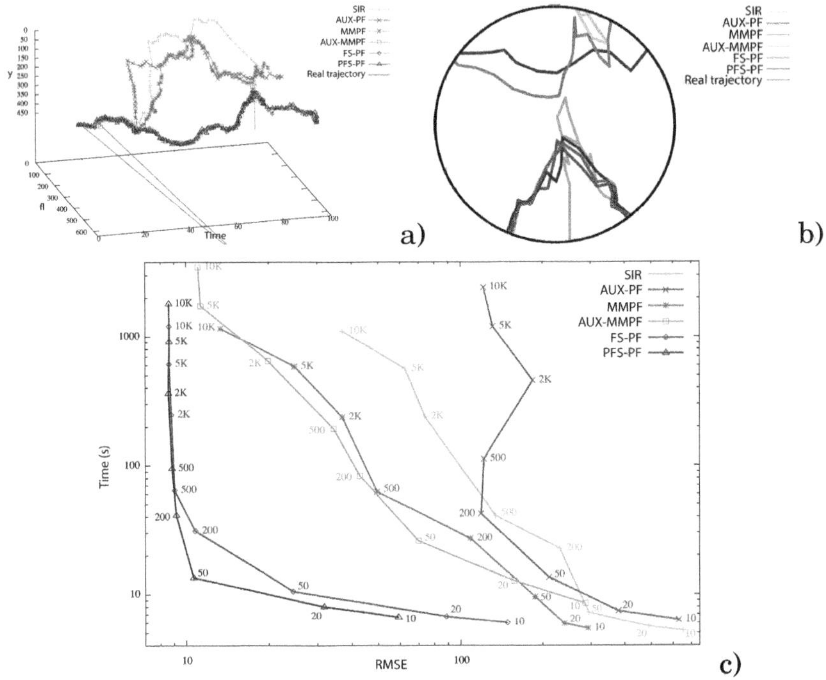

Figure 3.12. `Helicopter` *sequence*[6]*: a) trajectories estimated by different filters on a throw with* $N = 50$ *particles, b) close-up between* $t = 64$ *and* $t = 78$*, and c) computation costs (in seconds) as a function of the root mean square errors obtained by the different filters (the tags correspond to the number* N *of used particles). The two axes are logarithmic. For a color version of the figure, see www.iste.co.uk / dubuisson / tracking.zip*

For the first example, we consider a sequence containing 4 to 8 ants (the appearances and the disappearances of the ants are known in advance). The state of the object m,

6 http://www.kinetrack.org/experimental-data.html.

$\mathbf{x}_t^m = (x_t^m, y_t^m, \theta_t^m)^T$, contains its $2D$ position $(x_t^m, y_t^m)^T$ and its orientation θ_t^m. The position and orientation dynamics are random steps. The dimensions of the ant rectangles are set beforehand. The likelihoods are obtained with a technique of background subtraction, and correspond to the probabilistic modeling of foreground pixels. For this experiment, we used a spatial exclusion constraint by using a fuzzy semi-trapeze ensuring that two ants cannot superimpose more than 10% of their areas. Figure 3.13 gives a comparison of tracking results obtained with or without the integration of fuzzy spatial constraints. We can see that the addition of the spatial exclusion constraint makes it possible to lift the ambiguities introduced by the proximity of ants during certain time intervals, and that all the ants are tracked correctly.

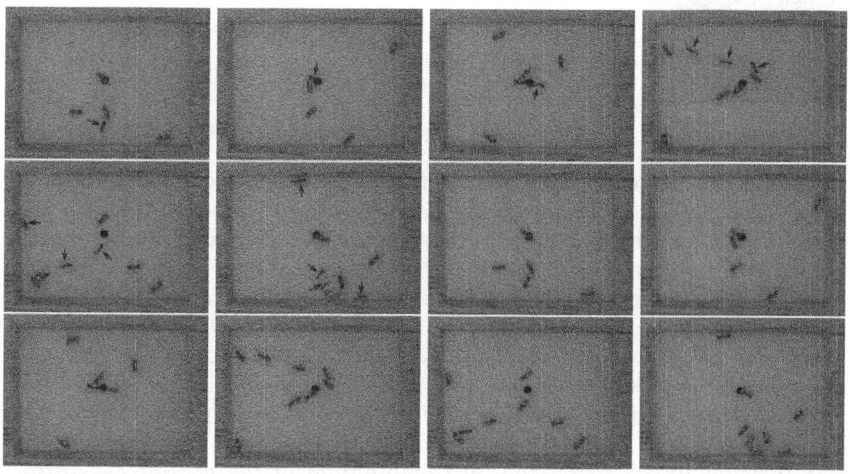

Figure 3.13. Ants *sequence. Results of tracking ants, obtained: top row, without the use of fuzzy spatial constraints (the blue arrows indicate that the ants are no longer tracked) and, bottom row, with their use. For a color version of the figure, see www.iste.co.uk/dubuisson/tracking.zip*

The second example concerns tracking four fingers of a hand, which can be seen as an articulated object. The state of the object m is modeled as previously. Every shape of the finger (assumed rigid) is set and represented by a *B-spline* curve, defined by a vector of six $2D$ control points. The position and orientation dynamics are random steps. The likelihood function is the one suggested in [ISA 98a], which uses the values of the maximal gradients on the normal lines of the *B-spline* control points. Four spatial constraints are implemented: distance (*Close* and *Far*), angle (*About* $\pi/8$ and *About* $-\pi/8$), alignment (the fingers should not deviate too much from the result of a linear regression) and exclusion (two fingers cannot superimpose by more than 50%), knowing that all of the constraints need to be satisfied and hence they are combined in a fuzzy conjunctive fashion. Figure 3.14 gives an illustration of the fuzzy spatial constraints used in the image space.

a) b) c) d) e)

Figure 3.14. *Sequence* `Main`. *Illustration of fuzzy spatial constraints in the image space (a darker gray level indicates stronger membership) for the case of the third finger. a) Fuzzy spatial constraint of the exclusion concept: neither the second, nor the fourth, not the fifth finger. b) Fuzzy spatial constraint of the angle concept, result constrained by the values of the surrounding fingers (second finger $-\pi/8$, fourth $\pi/8$ and fifth $\pi/8$). c) Fuzzy spatial constraint of the concept of distance, result constrained by the values of the surrounding fingers (second finger Near by, fourth Near by and fifth Far). d) Fuzzy spatial constraint of the alignment concept: the positions of the second, fourth and fifth fingers make it possible to construct a fuzzy line through linear regression. e) Combination of the spatial constraints illustrated in (a), (b), (c) and (d). For a color version of the figure, see www.iste.co.uk/dubuisson/tracking.zip*

Here again, the obtained results of the tracking were compared with and without the integration of fuzzy spatial constraints, as shown by the results in Figure 3.15. We can once more see the interest in adding fuzzy spatial constraints to the tracking process, as all of the fingers are tracked correctly, even when they are partially invisible.

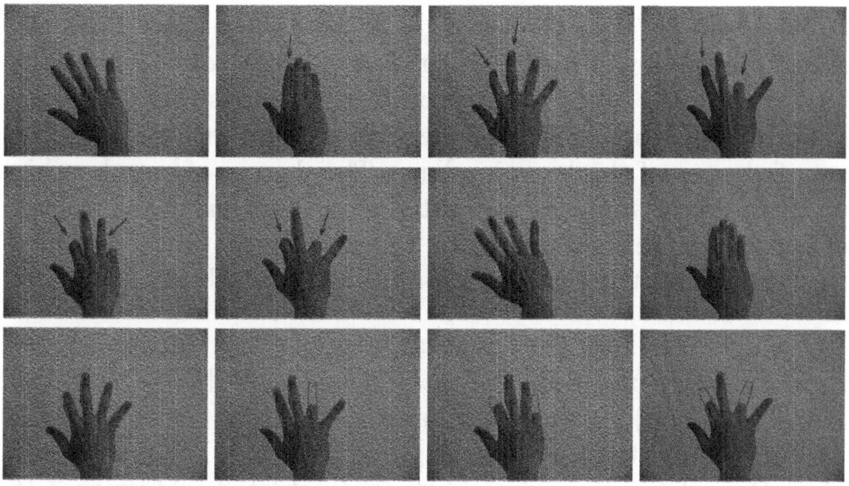

Figure 3.15. *Sequence* `Main`. *Results of tracking fingers, obtained at instants 100, 258, 432, 529, 620, 685 and 740, the top row does not use fuzzy spatial constraints (the blue arrows indicate incorrectly tracked fingers), and the bottom row does use them*

3.3.3.3. *Application to tracking shapes*

In this section, we take interest in the information of fuzzy shapes. In this context, a fuzzy set describes a set of shapes of varying degrees of membership, but grouped under a common tag. We call this tag form, that is, a fuzzy set characterizing a set of shapes: hence, every element of the fuzzy sets is a shape. Other approaches in the literature [PHA 01a, PHA 01b, TAN 09] suggested a fuzzy version of a model describing shapes. Nevertheless, using description models of mathematical shapes can be constraining, in the sense that the generalization of the approach to other types of shapes can be compromised. Moreover, these models usually do not integrate dynamic information, which is necessary to our application. As we see in this section, this approach is abstracted from all explicit mathematic modeling of shapes, as they are defined implicitly through constraints, by using a notion of distance between the candidate shape and a shape belonging to the fuzzy set under consideration.

We consider that the creation of fuzzy sets describing the shapes is achieved by learning. Every shape of the learning base contains a set of control points and a class tag. The fuzzy sets that are then implicitly generated take into account the set of shapes of the learning base. The degree of membership of a candidate shape is computed as a function of the degree of similarity, defined according to a chosen distance, between this shape and the closest shape from the database.

We give below the general scheme for learning and constructing fuzzy sets, for more details, see [WID 10a]. Let $BD = \{(a^d, c^d)\}_{d=1}^{D}$ be a database of elements $a^d \in \mathbb{R}^{2P}$ of P $2D$ control points assigned a tag $c^d \in \{1, \ldots, K\}$. The learning algorithm of the fuzzy functions is illustrated in Figure 3.16 (for example, by taking into account the shapes of the moth) is summarized by the following points:

1) Creation of a database:

- collection of D elements $\{(a^d, c^d)\}_{d=1}^{D}$;

- rotation of all the shapes in order to align their primary axes;

- reparation of elements by class.

2) Creation of fuzzy functions associated with elements (see [WID 10a])

3) Creation of fuzzy class functions $\{\mu_k^r\}_{k=1}^{K^r}$

The shape of the mouths is described by $P = 18$ control points $\{(x^p, y^p)\}_{p=1}^{P}$, divided between specific points of their contour (for this, we use the labeled template `Talking Face`). The family of the mouth form functions is obtained automatically by learning from these sets of points, giving elements a^d of BD, and three classes were considered: *Closed*,

Open and *Smile*. A mouth needs to be open to be able to form a smile. Therefore, when the lower and upper lips are not in contact, the considered class is either *Open* or *Smile*.

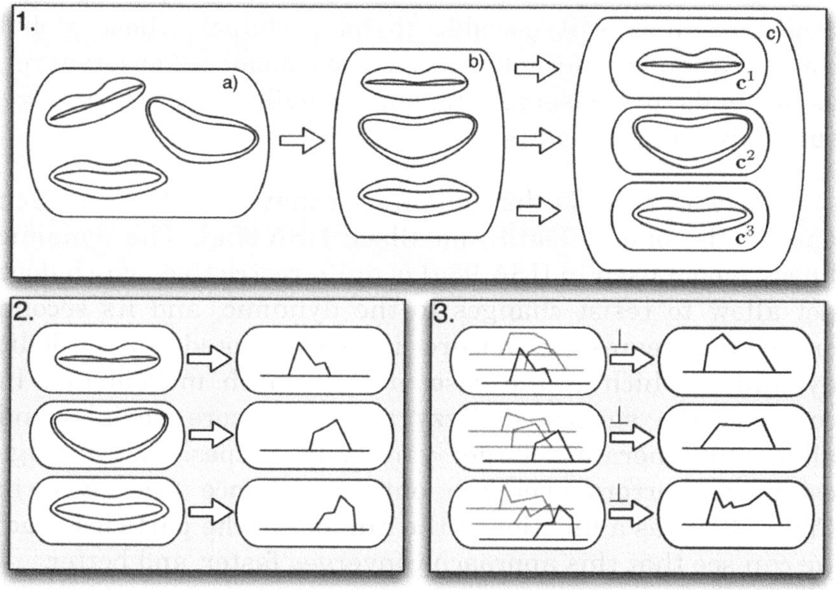

Figure 3.16. *Learning scheme for fuzzy functions (extract from [WID 10a]). The shapes used in this example are mouths. Stage 1 corresponds to the creation of the database, stage 2 to the creation of fuzzy functions associated with elements, and stage 3 to the creation of fuzzy class functions using a t-conorm* $\perp = \max$ *(fuzzy disjunction). In the stages 1 and 2, only one shape occurrence by class is represented, for the sake of clarity. The fuzzy sets are represented in dimension 1, even though in reality, they are defined over* $\mathbb{R}^{2P}$*, with* P *the number of control points*

In order to guarantee a global shape nearly identical to that of an element a^d, while allowing for changes in scale and local modifications, we define the shape distance as:

$$d(x, y) = \frac{1}{P}\sum_{p=1}^{P}\left|(\widehat{x^{p-1}, x^p, x^{p+1}}) - (\widehat{y^{p-1}, y^p, y^{p+1}})\right|$$

with $\widehat{(a,b,c)}$ the angle between vectors $\vec{ab}$ and $\vec{bc}$ and, by convention, $x^0 = x^P$, $y^0 = y^P$, $x^{P+1} = x^1$ and $y^{P+1} = y^1$. Even if the changes in scale and the local modifications remain minor, as they are constraint by fuzzy modeling, this shape constraint makes it possible to have shapes whose global configuration is coherent *vis-à-vis* the considered observation, as it tends to conserve the angles defined between three successive control points.

Figure 3.17 shows the contours estimated by this approach and by the one of Isard and Black [ISA 98a]. The dynamic linear model used in [ISA 98a] is quite restrictive, which does not allow to resist changes to the dynamic, and its second order auto-regressive nature is not adapted for periodic dynamics, which is the case of the mouth movements. In contrast, the suggested fuzzy model is more flexible, and allows to generate a larger variety of shapes. The average estimation errors over the entire sequence are given in Figure 3.18, as a function of the number of the particles used: we can see that this approach converges faster, and better.

Another implicit characteristic of our algorithm is that it is capable of recognizing the form of a mouth due to its learning from a database. The obtained *a posteriori* probabilities of the forms are given in Figure 3.20: the form having the highest *a posteriori* probability will be the one estimated. The comparison with the ground truth (given below the abscissa axis) shows that our approach estimates, on the whole, the correct forms. Although it is possible to observe a slight increase in values, the forms estimated in images 150 and 160 are not well-estimated, as this period corresponds to the times when the subject opens the mouth slightly. The ground truth does not integrate information on the degree of 'openness' of the mouth, which explains this result. Finally, the Monte-Carlo expectancy obtained in our test sequence, as well as the estimation of the form indices, are illustrated in Figure 3.19, by using $N = 5000$ particles.

Figure 3.17. `Talking Faces` *sequence*[7]*. contour estimations obtained by using $N = 50$ particles a) in the approach that learns fuzzy shapes and b) by the method of Isard et al. [ISA 98a]. For a color version of the figure, see www.iste.co.uk/dubuisson/tracking.zip*

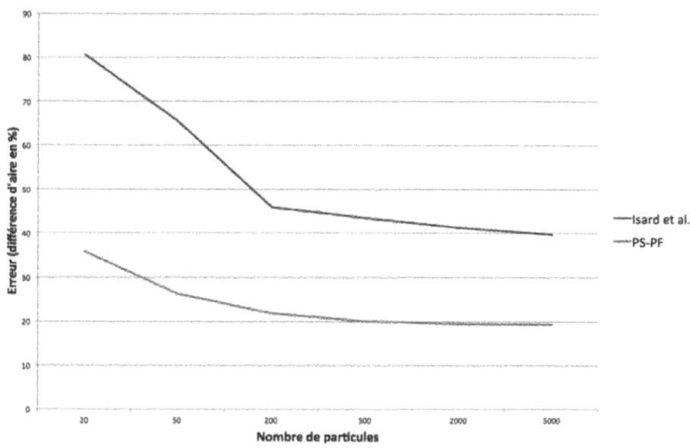

Figure 3.18. `Talking Faces` *sequence. Comparison of the average estimation errors between the approach of Isard et al. [ISA 98a] (in blue) and our approach (PS-PF, in red)*

7 http://personalpages.manchester.ac.uk/staff/timothy.f.cootes/data/talking_face/talking_face.html.

3.4. Conjoint estimation of dynamic and static parameters

In the work suggested in [HER 14], we are interested in estimating the shape of a face modeled by a dense $3D$ mesh, seen from several view-points at once. This $3D$ mesh, known under the name of $3D$ Morphable Model ($3DMM$) [BLA 99b], examples of which can be seen in Figure 3.20, is defined by many parameters describing its form, texture and shape, hence giving its representation in a large state space.

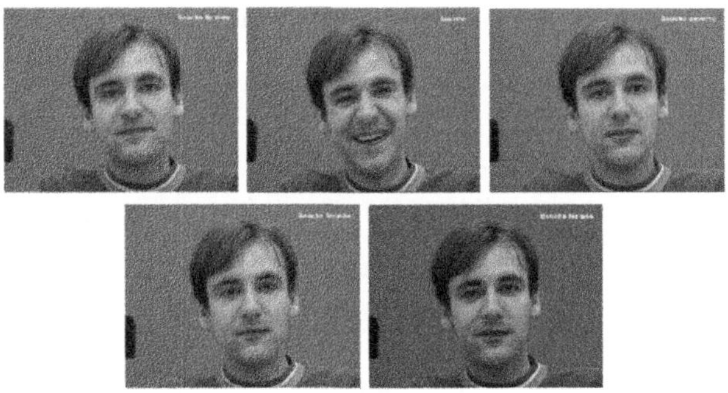

Figure 3.19. *`Talking Faces` sequence. Estimations of the contours and the postures obtained by out approach for $N = 5000$. For a color version of the figure, see www.iste.co.uk/dubuisson/tracking.zip*

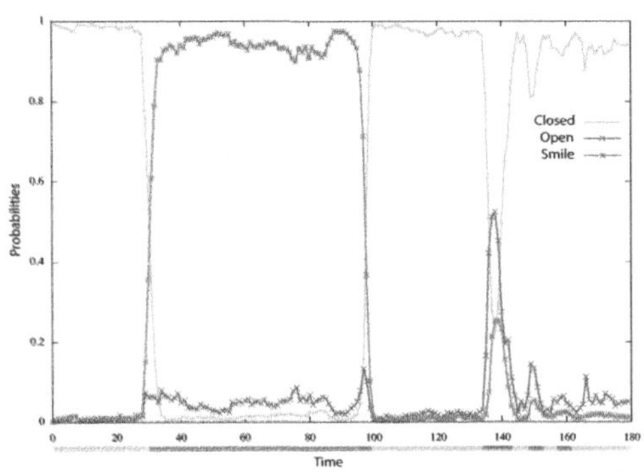

Figure 3.20. *`Talking Faces` sequence. a posteriori probabilities, with $N = 5000$ particles and in 180 images obtained by our approach. The ground truth is indicated under the abscissa axis. For a color version of the figure, see www.iste.co.uk/dubuisson/tracking.zip*

Most of the algorithms suggested to estimate parameters such as $3D$ models most often use a single image on the input and therefore are very dependent on the quality of the observations [BLA 99c, ROM 05]. However, to obtain the best results, it is more interesting to use several images in order to consolidate the temporal reconstruction. In [AMB 07], the authors suggest fusing the images by using a stereo vision technique. The use of video sequences has also been widely exploited, with the exemption of approaches of the "structure from motion" type, in which the images are considered as a set [FAG 08]. In the [ROO 11] works, the authors extend an approach working on a single image in the entire sequence by fusing the information obtained at every instant independently, without, however, ensuring the coherence of the model.

A common way of proceeding is to fit this $3D$ model at every instance onto the available observations by optimizing the form, shape and texture parameters. Thus, starting from an initialization, we are looking to minimize the cost function along several criteria. Among the optimization methods used in this context, we can mention the Levenberg-Marquardt algorithm [LEV 44, MAR 63], which is a non-linear optimization approach that consists of recursively minimizing error by gradient descent. The problem with this type of iterative method is that it is very dependent on the initialization and there is a risk of only reaching a local minimum far from the global minimum, in particular if the observations are noisy. Very few temporal fusion techniques were suggested [VAN 11], to estimate at every instant the shape of the model, by taking into time, and therefore the past estimations, into account.

Here, works suggest to temporarily estimate the $3D$ shape with a particle filter to update recursively the parameters of the model, rather than begin with an initialization at each instant [HER 12]. The geometry of the $3DMM$ model is learned through Principal Component Analysis (PCA) [BIS 07], which provides a shape model defined by a linear combination of deformations from which are computed the $3D$ coordinates of the vertices of the mesh, connected by edges and associated with faces. We call $\Theta = \{\theta^1, \ldots, \theta^D\}$ the vector containing these deformations, which follows the probability distribution $p(\Theta) = e^{-\frac{1}{2}\sum_{d=1}^{D}\left(\frac{\theta^d}{\lambda^d}\right)^2}$, where λ^d is the dth proper value of the variance-covariance matrix of the learning set. The problem is that the shape of the head influences its form (for example, if a face that is too thin is estimated in relation to the reality, it will not be possible to estimate correctly the form, as the position of the ears will be incorrect). It is therefore necessary to estimate conjointly the shape and form parameters at every instant, which increases the size of the state space. As a solution, we suggest introducing static variables into the particle filter, of the form $\Theta = \{\theta^1, \ldots, \theta^D\}$ and to use them to estimate the state vector $\mathbf{x}_t$ corresponding to the form. Thus, equation [1.1] defined in Chapter 1 becomes:

$$\begin{cases} \mathbf{x}_t = f_t(\mathbf{x}_{t-1}, \mathbf{u}_t) \\ \mathbf{y}_t = g_t(\mathbf{x}_t, \Theta, \mathbf{v}_t) \end{cases} \qquad [3.10]$$

As we can see, the transition does not depend on the deformation parameters as they do not influence the dynamic of the form. Inversely, the equation depends on the shape and the form, as both are used for correction.

Here, we could estimate the optimal deformation vector Θ^* directly, through optimization techniques, but we preferred integrating these parameters into the (hidden) state vector in order to estimate the conjoint density $p(\mathbf{x}_t, \Theta|\mathbf{y}_{1:t})$. Hence, every particle will represent simultaneously the shape and the form of the $3D$ model. There are very few works on this subject; to our knowledge, the only one that was applied to estimating the static parameters of simple geometric $3D$ shapes is suggested in [MIN 10]. In our case, the shapes are a lot more complex. In order to avoid the problem of degeneration, it is nevertheless preferable to add artificial movement to the static part of the particles to explore the space of these parameters. However, by definition, these parts are static and so the dynamic law that governs their evolution between two instances is the identity function. In our case, the movement of the static part of the particles is generated during the MCMC stage (Monte Carlo Markov Chain) [FEA 02] (see [HER 12] for more detail). The correction is made with the help of a likelihood function that is based on a combination of distances to the characteristic points and to the internal contours and on a measurement of similarity between silhouettes [HER 12]. The general functioning of the approach is given in the Algorithm 3.4.

Figure 3.21 gives the tracking results obtained in the three images of a sequence acquired in outside condition with three simultaneous views. For each view, we can see, projected on the image, on the left the average starting model, and on the right the model estimated by the particle filter (best particle). We can note that the estimated model is close enough to the average model at the start, but becomes closer to the shape of the observed face over time, showing the advantage of recursive temporal filtering. Figure 3.22 finally shows the comparative results of the estimation obtained over two sequences with four simultaneous views (acquired inside) with our approach and the Levenberg-Marquardt algorithm [LEV 44, MAR 63]. We can see in these two examples that the $3D$ position is on average better estimated in the four images of each sequence by particle filter.

The unimodal distribution hypothesis not being guaranteed, one of the points to improve is the characterization of the output state of the filter at every instant, from a set of particles. We have used the state given by the best particle, but we noticed that sometimes the mesh validates the observations in several images, even though the parameters associated with different instants are not similar. Using the best particle to characterize the optimal parameters is therefore not the optimal choice: mode detection would certainly allow to characterize the local maxima better.

Algorithm 3.4: Algorithm for conjoint estimation of static and dynamic parameters [HER 12]

Input : set of particles $\left\{\{\mathbf{x}_{t-1}^{(i)}, \Theta_{t-1}^{(i)}\}, w_{t-1}^{(i)}\right\}_{i=1}^{N}$

Output : set of particles $\left\{\{\mathbf{x}_{t}^{(i)}, \Theta_{t}^{(i)}\}, w_{t}^{(i)}\right\}_{i=1}^{N}$

$\left\{\{\mathbf{x}_{t-1}^{(i)}, \Theta_{t-1}^{(i)}\}, 1/N\right\}_{i=1}^{N} \leftarrow$ multinomial resampling of $\left\{\{\mathbf{x}_{t-1}^{(i)}, \Theta_{t-1}^{(i)}\}, w_{t-1}^{(i)}\right\}_{i=1}^{N}$

for $i = 1$ **to** N **do**

  Propagation: $\mathbf{x}_t^{(i)} \sim p\left(\mathbf{x}_t|\mathbf{x}_{t-1}^{(i)}\right)$, $\Theta_t^{(i)} = \Theta_{t-1}^{(i)}$

  Correction: $w_t^{(i)} \propto w_t^{(i)} p\left(\mathbf{y}_t|\mathbf{x}_t^{(i)}, \Theta_t^{(i)}\right)$

Normalization: $\left\{w_t^{(i)}\right\}_{i=1}^{N} \leftarrow \left\{\frac{w_t^{(i)}}{\sum_{j=1}^{N} w_t^{(j)}}\right\}_{i=1}^{N}$

(Optional) MCMC displacement on the statistical parameters $\Theta_t^{(i)}$

View 1 View 2 View 3

Figure 3.21. *Results of the 3D positions obtained in the sequence acquired outside with three simultaneous views. For each view, we present, projected, on the left the initial average model (previous estimation) and on the right, the estimated model (best particle). For a color version of the figure, see www.iste.co.uk/dubuisson/tracking.zip*

Figure 3.22. *Comparison of results obtained by our approach using a particle filter (blue model) and by the Levenberg-Marquardt algorithm (green model). Both sequences were tested, each containing for simultaneous views. For a color version of the figure, see www.iste.co.uk / dubuisson / tracking.zip*

3.5. Conclusion

In this chapter we suggested contributions allowing to reduce the search region in the state space, to focus the computations and therefore limit their complexity.

Two contributions were dedicated to multi-object tracking and integrate an association module between measurements and objects. A first approach aims to model a proposal function in order to propagate particles only towards moving areas. At every instant, a particle is associated with a single object following a classification that is updated as a function of the temporal context, which makes it possible to know the number of objects. This simple approach presents many advantages. The first is that it is not necessary to know the either appearance of the objects or their number to follow them: only movement is used, which makes it possible to simply manage the changes in illumination, to work by using the same algorithm with a set of objects whose number can vary over time, and, most importantly, to manage the cases of appearance and disappearance. The primary limit of this approach is that it assumes that all detected movement comes from an object (which is not necessarily the case), and that it detects stopping and disappearing objects in the same way, which leads to temporal non-linearity of the trajectories. The second approach allows to compute the association probabilities between the measurements and the objects by only using the past positions of the objects. An original geometric formulation makes it possible to lift the ambiguous association cases, to adapt to various movements (linear, oscillatory, erratic, etc.) and to surpass several state of the art data association filters. A significant improvement of computation time can also be considered: to compute the intersections between surfaces, we need to use optimized algorithms, such as those of `Numerical Recipes` [PRE 07], or libraries like `Boost`, which make the approach even more competitive.

Another presented solution consists of integrating fuzzy spatial information into the particle filter in order to add structural information about or between objects, which has not been done until now. The approach suggested is generic, as it allows to take into account different types of information. Thus, we have shown that it can be used for tracking single and multiple objects, or shapes, and that it gives better tracking results in the case of sudden movements. One of the disadvantages of this approach is that the suggested modeling for the importance function can make it costly if it is not constructed with carefully, that is, by taking care to add optimization stages such that they do not degrade the results. Moreover, an independence hypothesis, which can not be adapted to every situation, is made between the dynamic parameters. Finally, it is necessary to introduce more complex spatial relations, such as *Between*, which was for example suggested in [BLO 06].

Lastly, we presented preliminary works on estimating the $3D$ shape of a face with the help of a particle filter, by integrating dynamic and static parameters into the state vector. This original approach aims at solving the problems encountered by classical probabilistic deterministic methods, which estimate the current state (by maximum *a posteriori* or likelihood) as a function of the past state, through exhaustive or local searches (of the gradient descent type). The first results, on synthetic and real data, seem promising when compared to those obtained by an optimization technique. Nevertheless, they remain to be deeply analyzed.

The following chapter presents another way of tackling large dimensions of observation and state spaces, which consists of decomposing them into subspaces where subcomputations are possible in practice.

4

Models of Tracking by Decomposition of the State Space

4.1. Introduction

One of the major problems with using importance sampling in the particle filter (PF) described in Chapter 1 is the "curse of dimensionality" [MAC 98, ROB 05], that is the fact that the number of particles necessary for tracking increases exponentially with the size of the space and therefore the size of the chosen object description or the number of objects to follow, for example. The immediate consequence is that the computations to make in this space are no longer realizable once the description becomes precise or several objects need to be tracked. By exploiting the independence of the state and observation vector variables, it is possible to decompose their spaces into sub-spaces in such a manner that sub-computations over them would be realizable.

Among the suggested approaches, of which several are described in this chapter, partitioned sampling (PS) [MAC 00b] is undoubtedly the most popular approach; the latter exploits the decomposition of likelihood and transition densities into spaces of smaller dimensions. Hence, the state space $\mathcal{X}$ and observation space $\mathcal{Y}$ are decomposed in the following way: $\mathcal{X} = \mathcal{X}^1 \times \cdots \times \mathcal{X}^M$ and $\mathcal{Y} = \mathcal{Y}^1 \times \cdots \times \mathcal{Y}^M$. Exploiting the conditional independences between the sub-spaces leads to applying the PF algorithm not a single time over $(\mathcal{X}, \mathcal{Y})$, but sequentially over each of $(\mathcal{X}^m, \mathcal{Y}^m)$, $m = 1, \ldots, M$, which accelerates considerably the processing time. For instance, if we want to estimate the state of the articulated object represented in the left image of Figure 4.3, we would first estimate part 1, then part 2 given part 1, etc., until part 6 given all the previously estimated parts (for multiple objects cases, we proceed in the same way, object by object).

The general idea of PS is the following. We consider a state $\mathbf{x}_t$ decomposed into parts, or components, $\{\mathbf{x}_t^1, \ldots, \mathbf{x}_t^M\}$. The filtering distribution is approximated by the sample $\left\{\{\mathbf{x}_t^{1,(i)}, \ldots, \mathbf{x}_t^{M,(i)}\}, w_t^{(i)}\right\}_{i=1}^N$. The idea is to estimate this distribution part by part. If we consider the part m (object, part of an object, etc.), then we propagate the sub-particles $\mathbf{x}_t^{m,(i)}$ via a proposal function that depends on their previous state and the state of all the other sub-parts that have already been processed (depending on the case, the current observation could also be used in the proposal function, see Chapter 1). We define an m-tuple $\langle \mathbf{x}_{t-1}^{m,(i)}, \mathbf{x}_t^{1:m-1,(i)} \rangle$, which represents part m of particle i at instant $t-1$, as well as parts 1 to $m-1$ of the same particle at instant t (i.e. the parts that have already been processed). Weighted resampling (see section 1.2.3.5) then makes it possible to keep only the best m-tuples $\langle \mathbf{x}_{t-1}^{m,(i)}, \mathbf{x}_t^{1:m-1,(i)} \rangle$, i.e. the most representative sub-samples of the joint distribution of parts 1 to m. The weights thus obtained are then normalized. Once all of the parts have been processed, the complete weights of the particles are calculated as the product of the weights part by part. We give the general workings of the PS in Algorithm 4.1. Note that under the independence hypothesis between the parts $\mathbf{x}_t^m$, the transition and the proposal functions used in Algorithm 4.1 can be written in a simpler manner, as we have, respectively:

$$p\left(\mathbf{x}_t^{m,(i)}|\langle\mathbf{x}_{t-1}^{m,(i)},\mathbf{x}_t^{1:m-1,(i)}\rangle\right)=p\left(\mathbf{x}_t^{m,(i)}|\mathbf{x}_{t-1}^{m,(i)}\right)$$

and

$$q\left(\mathbf{x}_t^{m}|\langle\mathbf{x}_{t-1}^{m,(i)},\mathbf{x}_t^{1:m-1,(i)}\rangle,\mathbf{y}_{1:t}\right)=q\left(\mathbf{x}_t^{m}|\mathbf{x}_{t-1}^{m,(i)},\mathbf{y}_{1:t}\right).$$

Algorithm 4.1: Algorithm for the partition sampling particle filter, PS

Input : set of particles $\{\mathbf{x}_{t-1}^{(i)},w_{t-1}^{(i)}\}_{i=1}^N$
Output : set of particles $\{\mathbf{x}_{t}^{(i)},w_{t}^{(i)}\}_{i=1}^N$
$\left\{\mathbf{x}_{t-1}^{(i)},1/N\right\}_{i=1}^N \leftarrow$ multinomial resampling of $\left\{\mathbf{x}_{t-1}^{(i)},w_{t-1}^{(i)}\right\}_{i=1}^N$
for $m=1$ **to** M **do**
  for $i=1$ **to** N **do**
    Propagation: $\mathbf{x}_t^{m,(i)}\sim q\left(\mathbf{x}_t^{m}|\langle\mathbf{x}_{t-1}^{m,(i)},\mathbf{x}_t^{1:m-1,(i)}\rangle,\mathbf{y}_{1:t}\right)$
  $\left\{\langle\mathbf{x}_t^{1:m,(i)},\mathbf{x}_{t-1}^{m+1:M,(i)}\rangle,w_t^{(i)}/\rho_t^{(i)}\right\}_{i=1}^N \leftarrow$ weighted resampling of $\left\{\langle\mathbf{x}_t^{1:m,(i)},\mathbf{x}_{t-1}^{m+1:M,(i)}\rangle,w_t^{(i)},\rho_t^{(i)}\right\}_{i=1}^N$
  Normalization: $\tilde{w}_t^{(i)}=\frac{w_t^{(i)}}{\sum_{j=1}^N w_t^{(j)}}$
Correction:

$$w_t^{(i)}\propto\tilde{w}_t^{(i)}\frac{\prod_{m=1}^M p\left(\mathbf{y}_t^m|\mathbf{x}_t^{m,(i)}\right)p\left(\mathbf{x}_t^{m,(i)}|\langle\mathbf{x}_{t-1}^{m,(i)},\mathbf{x}_t^{1:m-1,(i)}\rangle\right)}{q\left(\mathbf{x}_t^{m,(i)}|\langle\mathbf{x}_{t-1}^{m,(i)},\mathbf{x}_t^{1:m-1,(i)}\rangle,\mathbf{y}_{1:t}\right)}\quad i=1,\ldots,N$$

Normalization: $\left\{w_t^{(i)}\right\}_{i=1}^N\leftarrow\left\{\frac{w_t^{(i)}}{\sum_{j=1}^N w_t^{(j)}}\right\}_{i=1}^N$

These algorithms are at the heart of the approaches described in this chapter. Section 4.2 suggests a first improvement to the PS in the context of multi-object tracking in order to process in an optimal order the objects that we assume independent and thus improve tracking performance. This order is estimated at every instant. Section 4.3 then describes an approach aiming to permute sub-spaces to improve the quality of the filtering distribution estimate, and finally section 4.4 describes an approach based on this permutation idea, which makes it possible to resample the particles efficiently. The latter two approaches were applied to the case of tracking one or several articulated objects.

4.2. Ranked partitioned sampling

The principle of PS is to partition the state space into m elements, which can be parts of the object [DUF 09] or even a set of objects [MAC 00b]. The elements are processed in some order, which we call scenario, following the scheme described in Algorithm 4.1. However, as it has been shown in [SMI 04], using a predefined scenario can cause the impoverishment of particles (for example, if an element is hidden by another, starting the scenario by the former element will disrupt the tracking). Branched partitioned sampling (BPS) [MAC 99a] suggested adding a visibility vector to the state vector and thus recursively groups particles with a realization identical to this vector, which implicitly generates a hypotheses tree. However, this approach quickly becomes too expensive, as for M objects, it generates $M!$ hypotheses. Furthermore, the algorithm can propagate particles in sub-trees of the hypotheses tree whose visibility vector is incorrect, without ever being able to deal with particles of other hypotheses sub-trees. Dynamic PS uses a combination of distributions to represent the *a posteriori* distribution. Every component represents a specific order for processing objects. This strategy improves the results of PS as it decreases the impoverishment phenomenon, which is particularly notable during occlusions. Nevertheless, as for BPS, the particles evolve in different subsets, without interaction. This idea has also been drawn upon in [DUF 09] to fuse several characteristics of a single object. However, the number of particles allocated to each component is fixed, which can degrade the performance of the filter when the chosen orders are inappropriate [SMI 04].

To solve these problems, we suggested ranked PS (RPS) [WID 12], which estimates both the joint distribution and the order in which the components, or parts, are processed considering the set of possible permutations and implicitly pruning the inadequate scenarios. The components with the strongest confidences will be considered in the first stages of the scenario, as they are the ones most capable of overcome the particle impoverishment phenomenon.

In what follows, we consider the problem of multi-object tracking. For this reason, one component corresponds to one object and the whole state vector contains the parameters describing the set of objects. Let $\mathbf{o}_t = (\mathbf{o}_t^1, \ldots, \mathbf{o}_t^M)$ be a processing order, that is a permutation of M objects. The variable $\mathbf{o}_t^m$ indicates the position of object m in this sequence. We call the scenario reciprocal permutation $\mathbf{o}_t^{-1}$ considered at a particular instant t, which we denote as $\mathbf{s}_t = (\mathbf{s}_t^1, \ldots, \mathbf{s}_t^M)$. Hence, the kth component of a scenario is defined such that $\mathbf{s}_t^k = m$ if and only if $\mathbf{o}_t^m = k$. For example, consider four objects and a processing order such that $\mathbf{o} = (2, 3, 4, 1)$. This means that the object identified as 1 will be considered in 2nd position, object 2 in 3rd, etc. The scenario corresponds then to $\mathbf{s} = (4, 1, 2, 3)$ and indicates in an ordered fashion the identifiers of the objects in the sequential approximation process (object 4 in the first position, etc.).

We first consider the probabilities of position transition to be set:

$$\mathbb{P}(\mathbf{s}_t^h = m | \mathbf{s}_{t-1}^k = m) \triangleq \mathbb{P}(\mathbf{o}_t^m = h | \mathbf{o}_{t-1}^m = k) \\ \triangleq \alpha_{k,h} \quad \forall m \in \{1, \ldots, M\} \qquad [4.1]$$

Modeling the treatment order $\mathbf{o}_t$ is necessary, as we wish to model the transition $\mathbb{P}(\mathbf{o}_t^m = h | \mathbf{o}_{t-1}^m = k)$, which indicates the probability for the object m to pass from position k to position h and the vector $\mathbf{s}_t$ makes it possible to designate the index of the object to consider at instant k of the sequential estimation process.

The use of a fixed transition matrix allows to set intuitively the transition probabilities of the discrete process. For modeling the processing order transition matrix, we favor the static states; hence, the matrix determined by the coefficients $\alpha_{k,h}$ should have the highest values along its diagonal, without systematically being symmetrical. By first considering the objects ordered in the first stages at time $t-1$, the joint distribution of the transition of $\mathbf{o}_t$ is written as:

$$\mathbb{P}(\mathbf{o}_t|\mathbf{o}_{t-1}) = \mathbb{P}\left(\mathbf{o}_t^{s^1}|\mathbf{o}_{t-1}^{s^1}\right)\prod_{k=2}^{M}\mathbb{P}\left(\mathbf{o}_t^{s^k}|\mathbf{o}_{t-1}^{s^k},\mathbf{o}_t^{s^1},\ldots,\mathbf{o}_t^{s^{k-1}}\right) \qquad [4.2]$$

with $\mathrm{s}^k \triangleq \mathrm{s}_{t-1}^k$, the time index being omitted to simplify the notation.

Going back to the example with $\mathbf{o}=(2,3,4,1)$ and considering that this estimation took place at time $t-1$, equation [4.2] gives:

$$\mathbb{P}(\mathbf{o}_t|\mathbf{o}_{t-1}) = \mathbb{P}(\mathbf{o}_t^4|\mathbf{o}_{t-1}^4)\,\mathbb{P}(\mathbf{o}_t^1|\mathbf{o}_{t-1}^1,\mathbf{o}_t^4)\,\mathbb{P}(\mathbf{o}_t^2|\mathbf{o}_{t-1}^2,\mathbf{o}_t^4,\mathbf{o}_t^1)$$
$$\times\mathbb{P}(\mathbf{o}_t^3|\mathbf{o}_{t-1}^3,\mathbf{o}_t^4,\mathbf{o}_t^1,\mathbf{o}_t^2)$$

The decomposition of the joint distribution $\mathbb{P}(\mathbf{o}_t|\mathbf{o}_{t-1})$ is therefore determined by the estimate at $t-1$ of the scenario, $\mathrm{s}_{t-1} = (4,1,2,3)$.

The last conditional distribution in equation [4.2] depends on the transition probabilities of the positions defined in equation [4.1], as well as the now inaccessible positions of the objects that have already been considered:

$$\mathbb{P}\left(\mathbf{o}_t^{s^k}=h|\mathbf{o}_{t-1}^{s^k}=k,\mathbf{o}_t^{s^1},\ldots,\mathbf{o}_t^{s^{k-1}}\right) = \left[1-\sum_{j=1}^{k-1}\delta^h_{\mathbf{o}_t^{s^j}}\right] \qquad [4.3]$$
$$\times\left[\alpha_{k,h}+\frac{1}{M-k+1}\sum_{j=1}^{k-1}\alpha_{k,\mathbf{o}_t^{s^j}}\right]$$

We can show that this distribution is indeed a probability distribution [WID 12]. The first term in the product guarantees that the probability is null, if position h has already been assigned, and 1 otherwise. The second term contains the *a priori* transition probabilities of position h to k, as defined in equation [4.1], as well as a term that uses the transition probabilities of the assigned positions to balance the distribution (these are the positions whose objects have already been processed, $\mathbf{o}_t^{s^1}, \ldots, \mathbf{o}_t^{s^{k-1}}$), and redistributes their probabilities uniformly over the positions that have not been attributed yet (all those whose first term is 1). As $k-1$ positions have already been attributed, $M-(k-1)$ remain available, which explains the denominator of the factor in the last sum.

To simplify the understanding of equation [4.3], reconsider the example with $\mathbf{o}_{t-1} = (2,3,4,1)$ and $\mathbf{s}_{t-1} = (4,1,2,3)$. The equations are then written as:

$$\mathbb{P}(\mathbf{o}_t^4 = h | \mathbf{o}_{t-1}^4 = 1) = \alpha_{1,h}$$

$$\mathbb{P}(\mathbf{o}_t^1 = h | \mathbf{o}_{t-1}^1 = 2, \mathbf{o}_t^4) = \left[1 - \delta^h_{\mathbf{o}_t^4}\right]\left[\alpha_{2,h} + \frac{1}{3}\alpha_{2,\mathbf{o}_t^4}\right]$$

$$\mathbb{P}(\mathbf{o}_t^2 = h | \mathbf{o}_{t-1}^2 = 3, \mathbf{o}_t^4, \mathbf{o}_t^1) = \left[1 - (\delta^h_{\mathbf{o}_t^4} + \delta^h_{\mathbf{o}_t^1})\right] \times \left[\alpha_{3,h} + \frac{1}{2}(\alpha_{3,\mathbf{o}_t^4} + \alpha_{3,\mathbf{o}_t^1})\right]$$

$$\mathbb{P}(\mathbf{o}_t^3 = h | \mathbf{o}_{t-1}^3 = 4, \mathbf{o}_t^4, \mathbf{o}_t^1, \mathbf{o}_t^2) = \left[1 - (\delta^h_{\mathbf{o}_t^4} + \delta^h_{\mathbf{o}_t^1} + \delta^h_{\mathbf{o}_t^2})\right] \times \left[\alpha_{4,h} + \alpha_{4,\mathbf{o}_t^4} + \alpha_{4,\mathbf{o}_t^1} + \alpha_{4,\mathbf{o}_t^2}\right]$$

We use the following decomposition of the joint transition density $p(\mathbf{x}_t, \mathbf{o}_t | \mathbf{x}_{t-1}, \mathbf{o}_{t-1})$:

$$p(\mathbf{x}_t, \mathbf{o}_t | \mathbf{x}_{t-1}, \mathbf{o}_{t-1}) = p(\mathbf{x}_t | \mathbf{x}_{t-1}, \mathbf{o}_t)\, \mathbb{P}(\mathbf{o}_t | \mathbf{o}_{t-1}) \qquad [4.4]$$

with $p(\mathrm{o}_t|\mathrm{o}_{t-1})$ the transition law defined in equation [4.2]. Given the processing order defined by o_t, the transition density of the state vector x_t is deconstructed by considering first the objects placed in the first stages:

$$p(\mathrm{x}_t|\mathrm{x}_{t-1}, \mathrm{o}_t) \triangleq \prod_{k=1}^{M} p\left(\mathrm{x}_t^{s_t^k}|\mathrm{x}_{t-1}^{s_t^k}, \mathrm{x}_t^{s_t^1}, \ldots, \mathrm{x}_t^{s_t^{k-1}}\right) \qquad [4.5]$$

Hence, the order of the Markov chain defined by the components of x_t, i.e. the components of the objects $\{\mathrm{x}_t^m\}_{m=1}^M$, is determined by the discrete process of the processing order of objects o_t or, equivalently, by the discrete process representing a scenario s_t.

Reconsider the example of the $\mathrm{s}_{t-1} = (4, 1, 2, 3)$ scenario and imagine that its transition at t is such that $\mathrm{s}_t = (4, 2, 1, 3)$, then s_t defines the decomposition of the joint probability density of the state dynamic described in equation [4.5]:

$$p(\mathrm{x}_t|\mathrm{x}_{t-1}, \mathrm{o}_t) = p(\mathrm{x}_t^4|\mathrm{x}_{t-1}^4)\, p(\mathrm{x}_t^2|\mathrm{x}_{t-1}^2, \mathrm{x}_t^4)\, p(\mathrm{x}_t^1|\mathrm{x}_{t-1}^1, \mathrm{x}_t^4, \mathrm{x}_t^2)$$
$$\times p(\mathrm{x}_t^3|\mathrm{x}_{t-1}^3, \mathrm{x}_t^4, \mathrm{x}_t^2, \mathrm{x}_t^1)$$

We set the dynamic process such that:

$$p(\mathrm{x}_t^{s_t^k}|\mathrm{x}_{t-1}^{s_t^k}, \mathrm{x}_t^{s_t^1}, \ldots, \mathrm{x}_t^{s_t^{k-1}}) = p(\mathrm{x}_t^{s_t^k}|\mathrm{x}_{t-1}^{s_t^k}) \qquad [4.6]$$

which means that an object is only conditioned by itself, at the previous instant. Hence, the likelihood is written as:

$$p(\mathrm{y}_t|\mathrm{x}_t, \mathrm{o}_t) \triangleq \prod_{k=1}^{M} p(\mathrm{y}_t|\mathrm{x}_t^{s_t^k}) \qquad [4.7]$$

To summarize, for each particle, at instant t, the algorithm first generates the processing order of the object, and therefore a scenario. Next, at every stage k of the partitioning process, the algorithm resamples the particle cloud, before proposing a new state for the object designated by the scenario at position k, and then computes the likelihood. The approximation of the joint filter law of $(\mathbf{x}_t, \mathbf{o}_t)$ is obtained once the M operations have been executed. The approximation procedure is described in Algorithm 4.2.

Algorithm 4.2: Algorithm of the ranked partitioning filter, RPS [WID 12]

Input : set of particles $\{\mathbf{x}_{t-1}^{(i)}, w_{t-1}^{(i)}\}_{i=1}^N$

Output : set of particles $\{\mathbf{x}_t^{(i)}, w_t^{(i)}\}_{i=1}^N$

$\left\{\mathbf{x}_{t-1}^{(i)}, 1/N\right\}_{i=1}^N \leftarrow$ multinomial resampling of $\left\{\mathbf{x}_{t-1}^{(i)}, w_{t-1}^{(i)}\right\}_{i=1}^N$

Generate the object processing order and the scenario: **for** $i = 1, \ldots, N$ **do**

- Generate $\mathbf{o}_t^{(i)} \sim \mathbb{P}(\mathbf{o}_t | \mathbf{o}_{t-1}^{(i)})$ according to equation [4.2]
- Compute the scenario $\mathbf{s}_t^{(i)}$ such that $\forall m = 1, \ldots, M, \mathbf{s}_t^{m,(i)} = \sum_{k=1}^M k\, \delta^m_{\mathbf{o}_t^{k,(i)}}$

for $k = 1$ **to** M **do**

- **for** $i = 1$ **to** N **do**
  - Propagation: $\mathbf{x}_t^{\mathbf{s}_t^{k,(i)},(i)} \sim q\left(\mathbf{x}_t^{\mathbf{s}_t^{k,(i)}} | \langle \mathbf{x}_{t-1}^{\mathbf{s}_t^{k,(i)},(i)}, \mathbf{x}_t^{\mathbf{s}_t^{1,(i)}:\mathbf{s}_t^{k-1,(i)},(i)} \rangle, \mathbf{y}_{1:t}\right)$
- $\left\{\langle x_t^{s_t^{1,(i)}:s_t^{k,(i)},(i)}, x_{t-1}^{s_t^{k+1,(i)}:s_t^{M,(i)},(i)}, s_t^{(i)} \rangle, \frac{w_t^{(i)}}{\rho_t^{(i)}}\right\} \leftarrow$ weighted resampling of $\left\{\langle x_t^{s_t^{1,(i)}:s_t^{k,(i)},(i)}, x_{t-1}^{s_t^{k+1,(i)}:s_t^{M,(i)},(i)}, s_t^{(i)} \rangle, w_t^{(i)}, \rho_t^{(i)}\right\}$
- Normalization: $\tilde{w}_t^{(i)} = \frac{w_t^{(i)}}{\sum_{j=1}^N w_t^{(j)}}$

Correction:

$$w_t^{(i)} \propto \tilde{w}_t^{(i)} \frac{\prod_{k=1}^M p\left(\mathbf{y}_t^{\mathbf{s}_t^{k,(i)}} | \mathbf{x}_t^{\mathbf{s}_t^{k,(i)},(i)}\right) p\left(\mathbf{x}_t^{\mathbf{s}_t^{k,(i)},(i)} | \langle \mathbf{x}_{t-1}^{\mathbf{s}_t^{k,(i)},(i)}, \mathbf{x}_t^{\mathbf{s}_t^{1,(i)}:\mathbf{s}_t^{k-1,(i)},(i)} \rangle\right)}{q\left(\mathbf{x}_t^{\mathbf{s}_t^{k,(i)},(i)} | \langle \mathbf{x}_{t-1}^{\mathbf{s}_t^{k,(i)},(i)}, \mathbf{x}_t^{\mathbf{s}_t^{1,(i)}:\mathbf{s}_t^{k-1,(i)},(i)} \rangle, \mathbf{y}_{1:t}\right)} \qquad i = 1, \ldots, N$$

Normalization: $\left\{w_t^{(i)}\right\}_{i=1}^N \leftarrow \left\{\frac{w_t^{(i)}}{\sum_{j=1}^N w_t^{(j)}}\right\}_{i=1}^N$

This approach mitigates the dimensionality problem by proceeding sequentially to the waited resampling stages in the state spaces of the individually taken objects. However, contrary to the classical PS method, processing order is also estimated by the PF, which makes the tracking more robust and reduces the impoverishment phenomenon, while maintaining identical computational time to PS. Some of the results obtained by this approach and BPS are presented in Figure 4.1, where we can see that RPS manages occlusion problems better than BPS. Tracking errors obtained with the two approaches are given in Figure 4.2. A more detailed analysis, as well as the parameters used for the tracking, are provided in [WID 10b, WID 12].

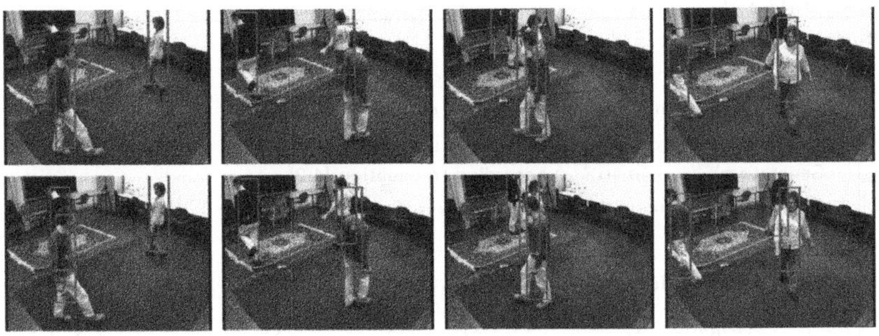

a)

b)

Image 24 Image 99 Image 205 Image 259

Figure 4.1. *Results of tracking four people in sequence* `Pedestrians` *[FLE 08]*[1] *obtained for $N = 500$ by a) BPS and b) RPS [WID 12]: RPS manages occlusions better, as, for example, in instants 24, 205 and 259. For a color version of the figure, see www.iste.co.uk/dubuisson/tracking.zip*

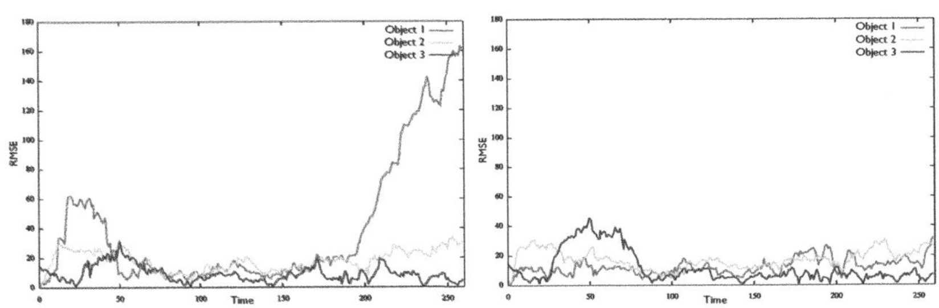

a) b)

Figure 4.2. *Tracking errors in sequence* `Pedestrians` *[FLE 08] obtained for $N = 500$ by a) BPS and b) RPS [WID 12]. For a color version of the figure, see www.iste.co.uk/dubuisson/tracking.zip*

1 http://cvlab.epfl.ch/research/body/surv/.

In the upcoming sections, we present a different point of view on taking into account large state spaces in the PF, by suggesting an approach that reduces the computation time and improves the estimation quality. This idea is then extended to the definition of a new resampling method.

4.3. Weighted partitioning with permutation of sub-particles

As we have seen, the key idea of PS and its variations consists in exploiting probabilistic independences in order to execute only PF computations within sub-spaces $(\mathcal{X}^m, \mathcal{Y}^m)$. Their dimension being "small", the conditional parameter distributions to estimate also have less parameters than the one defined over $(\mathcal{X}, \mathcal{Y})$, which reduces exponentially the number of particles necessary for their estimation and therefore accelerate the computations accordingly. Hence, if we apply the PS principle to the case of an articulated object in Figure 4.3, in order to, for example, follow the movements of a person, $\mathcal{X}$ decomposes naturally into $\mathcal{X}^{\text{torso}} \times \mathcal{X}^{\text{head}} \times \mathcal{X}^{\text{left arm}} \times \mathcal{X}^{\text{right arm}} \times \mathcal{X}^{\text{left forearm}} \times \mathcal{X}^{\text{right forearm}}$. If we know the position of the torso of the person to track, we can reasonably assume that there is independence between the positions of the head, the left arm and the right arm. Likewise, the positions of the left and right forearms are independent. This suggests a decomposition into the product of the conditional probabilities of the joint distribution over $(\mathcal{X}, \mathcal{Y})$ and over the set of time steps t, which is encoded naturally by the structure of a dynamic Bayesian network (DBN).

DEFINITION 4.1 (DBN [MUR 02]).– A Bayesian network of a couple $(G, \mathbf{P})$, where $G = (\mathbf{V}, \mathbf{A})$ is an acyclic directed graph. Every node $X \in \mathbf{V}$ corresponds to a random variable[2].

2 By abuse of notation, as there is a bijection between the nodes of $\mathbf{V}$ and the corresponding random variables, we use $X \in \mathbf{V}$ to represent indifferently the node X or its random variable.

$\mathbf{P} = \{p(X|\mathbf{Pa}(X)) : X \in \mathbf{V}\}$, where $\mathbf{Pa}(X)$ is the set of parents of X in G (we similarly define $\mathbf{Desc}(X)$ the set of its descendants), which is a set of conditional probabilities (or densities) and the joint probability $p(\mathbf{V})$ is given by $p(\mathbf{V}) = \prod_{X \in \mathbf{V}} p(X|\mathbf{Pa}(X))$. A DBN is an extension of a Bayesian network incorporating a notion of "temporarily": the nodes of $\mathbf{V}$ are then indexed by time: $\mathbf{V} = \cup_t\{\mathbf{x}_t, \mathbf{y}_t\}$.

Figure 4.3 gives an example of modeling the human body by a DBN: the torso corresponds to the variable $\mathbf{x}_t^1$, the head to $\mathbf{x}_t^6$, the left arm to $\mathbf{x}_t^2$, the right arm to $\mathbf{x}_t^4$, the left forearm to $\mathbf{x}_t^3$ and the right forearm to $\mathbf{x}_t^5$. The right side of the figure represents the graphical part of the DBN.

The DBN and the PF have already been used conjointly for tracking. In [KAN 95], the proposal function q is decomposed as the product of conditional densities in every node of the network and the algorithm of the PF is applied sequentially over these nodes according to a topological order in the DBN. These two approaches are combined in [ROS 08]. in order to integrate the algorithm of the PF into a DBN entirely, to define what can be seen as a generalization of the PS. Other methods founded on Bayesian networks were also developed to manage large-dimensional spaces. For example, the non-parametric belief propagation technique [SUD 10] or even the Rao-Blackwellised PF (RBPF) for DBN [DOU 00a], which decomposes the state space into two parts according to the strong hypothesis that the conditional distribution of the second part can be estimated by classical methods (Kalman filter, for example). Finally, we can mention the works in [ORT 02] that are closer to ours. In the applicative context of tracking M targets, with an independence hypothesis between these targets, these works suggest permuting the sub-samples, each associated with a target, in a spirit similar to the one described here. However, they differ primarily over two essential points. First of all, this permutation is executed differently: the new samples are formed by M weighted resamplings (one per target), executed independently. Next, and above all, the suggested approach cannot be generalized, as it is based on a strong hypothesis between the targets, which makes it inapplicable to our context. In the approach suggested in [DUB 14], only the conditional independences of the considered problem are exploited, which makes it possible to considerably extend its application field (although it makes proofs of validity significantly more complex).

The work described in this part exploits the independences, when they exist, of the DBN described by the d-separation property defined in [PEA 88]. If we return to the DBN in Figure 4.3, by d-separation, x_t^2, x_t^4 and x_t^6 are conditionally independent of $(x_t^1, x_{t-1}^2, x_{t-1}^4, x_{t-1}^6)$, which implies that in PS, the parts 2, 4 and 6 can be propagated/corrected in parallel, rather than sequentially, while guaranteeing that the estimated distribution is correct. Likewise, x_t^3 and x_t^5 are conditionally independent of, respectively, (x_t^2, x_{t-1}^3) and (x_t^4, x_{t-1}^5). The first advantage of proceeding this way is that we only need one resampling instead of three for filtering parts 2, 4 and 6. The second advantage is that it is possible to permute, among certain sets of particles, the sub-states of parts 2, 4 and 6, and then the sub-states of 3 and 5, in order to focus on the distribution modes to estimate without, however, modifying the estimated distribution itself. We do not give here the demonstrations, which are available in [DUB 12b]; however, we give the general principle behind these two ideas.

4.3.1. *Permutation of sub-samples*

We consider no longer the particles $\{x_t^{(i)}, w_t^{(i)}\}$ at time t, but their total trajectories from instant 0, or $\{x_{0:t}^{(i)}, w_{0:t}^{(i)}\}$. Within a normalization factor, they sample the distribution $p(x_{0:t}, y_{1:t})$. In the same way that the states x_t decompose into variables x_t^m, $m = 1, \ldots, M$, the weights $w_t^{(i)}$ decompose into

$w_t^{m,(i)} = p(\mathbf{y}_t^m|\mathbf{x}_t^m)$, i.e. the likelihood of the part m. A weight $w_t^{(i)}$ is then defined by $w_t^{(i)} \propto \prod_{m=1}^{M} w_t^{m,(i)}$ and $w_{0:t}^{(i)} = \prod_{s=0}^{t} w_s^{(i)}$. Then, $\{\mathbf{x}_{0:t}^{m,(i)}, w_{0:t}^{m,(i)}\}$ samples the distribution $p(\mathbf{x}_{0:t}^m, \mathbf{y}_{1:t}^m|\mathbf{Pa}(\mathbf{x}_{0:t}^m))$, where $\mathbf{Pa}(\mathbf{x}_{0:t}^m)$ represents $\bigcup_{s=0}^{t}\{$ parents of nodes $\mathbf{x}_s^m$ in the DBN $\}$. And:

$$p(\mathbf{x}_{0:t}, \mathbf{y}_{1:t}) = \prod_{m=1}^{M} p(\mathbf{x}_{0:t}^m, \mathbf{y}_{1:t}^m|\mathbf{Pa}(\mathbf{x}_{0:t}^m))$$

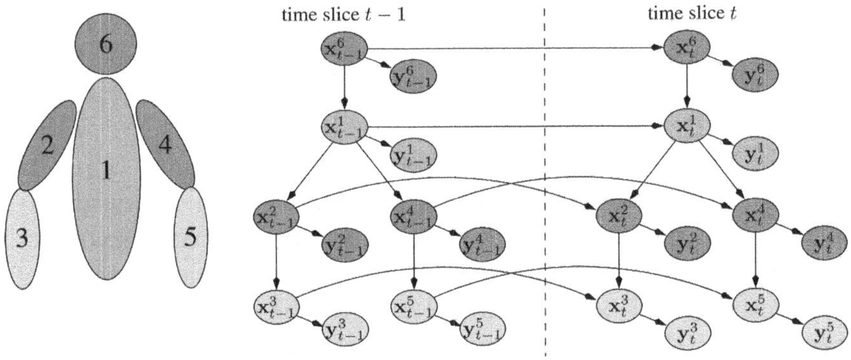

Figure 4.3. *Example illustrating the modeling of an articulated object (a human body, on the left) by a DBN (on the right) over two time intervals. In this schema, the part m at instant t corresponds to the random variable $\mathbf{x}_t^m$ encoded in a node of the DBN. For a color version of the figure, see www.iste.co.uk/dubuisson/tracking.zip*

For example, on the DBN of Figure 4.3, $\{\mathbf{x}_{0:t}^{2,(i)}, w_{0:t}^{2,(i)}\}$, i.e. the set of part 2 of the particles over all the time intervals, samples $p(\mathbf{x}_{0:t}^2, \mathbf{y}_{1:t}^2|\mathbf{x}_{0:t}^1)$. Likewise, $\{\mathbf{x}_{0:t}^{1,(i)}, w_{0:t}^{1,(i)}\}$ and $\{\mathbf{x}_{0:t}^{4,(i)}, w_{0:t}^{4,(i)}\}$ sample, respectively, $p(\mathbf{x}_{0:t}^1, \mathbf{y}_{1:t}^1)$ and $p(\mathbf{x}_{0:t}^4, \mathbf{y}_{1:t}^4|\mathbf{x}_{0:t}^1)$ and the joint law decomposes into the product of six distributions:

$$p(\mathbf{x}_{0:t}^1, \mathbf{y}_{1:t}^1) \times p(\mathbf{x}_{0:t}^2, \mathbf{y}_{1:t}^2|\mathbf{x}_{0:t}^1) \times p(\mathbf{x}_{0:t}^4, \mathbf{y}_{1:t}^4|\mathbf{x}_{0:t}^1) \times$$
$$p(\mathbf{x}_{0:t}^6, \mathbf{y}_{1:t}^6|\mathbf{x}_{0:t}^1) \times p(\mathbf{x}_{0:t}^3, \mathbf{y}_{1:t}^3|\mathbf{x}_{0:t}^2) \times p(\mathbf{x}_{0:t}^5, \mathbf{y}_{1:t}^5|\mathbf{x}_{0:t}^4)$$

Yet, by d separation, the set $\{\mathbf{x}^2_{0:t}, \mathbf{y}^2_{1:t}\}$ is independent of $\{\mathbf{x}^4_{0:t}, \mathbf{y}^4_{1:t}\}$ conditionally to $\{\mathbf{x}^1_{0:t}\}$. Hence, if we swap two "sub-particles" $\{\mathbf{x}^{4,(i)}_{0:t}, w^{4,(i)}_{0:t}\}$ and $\{\mathbf{x}^{4,(j)}_{0:t}, w^{4,(j)}_{0:t}\}$ such that $\{\mathbf{x}^{1,(i)}_{0:t}\} = \{\mathbf{x}^{1,(j)}_{0:t}\}$, this will have an impact neither on the estimation of $p(\mathbf{x}^2_{0:t}, \mathbf{y}^2_{1:t}|\mathbf{x}^1_{0:t})$ (by independence), nor on the one of $p(\mathbf{x}^4_{0:t}, \mathbf{y}^4_{1:t}|\mathbf{x}^1_{0:t})$ (as permuting the elements in a sample in no way affects the estimation that it provides). Even more so, the estimated joint distribution is unchanged. However, this operation makes it possible to focus on the modes of this distribution. Indeed, consider Figure 4.4, where three particles $(\mathbf{x}^{(1)}_{0:t}, \mathbf{x}^{(2)}_{0:t}, \mathbf{x}^{(3)}_{0:t})$ are represetned and $\mathbf{x}^1_{0:t}, \mathbf{x}^2_{0:t}, \mathbf{x}^4_{0:t}$ of Figure 4.3 being, respectively, the central $(\{\mathbf{x}^{1,(i)}_{0:t}\}_{i=1,2,3})$, the left $(\{\mathbf{x}^{2,(i)}_{0:t}\}_{i=1,2,3})$ and the right $(\{\mathbf{x}^{4,(i)}_{0:t}\}_{i=1,2,3})$ parts of every particle in this figure (the gray parts representing the real state of the object). The first two particles have the same value for $\mathbf{x}^1_{0:t}$, and we can swap their $\mathbf{x}^2_{0:t}$ and $\mathbf{x}^4_{0:t}$ values. Note that before a permutation (top row), the right part (respectively, left) of the first (respectively, second) particle was too far from the real state and therefore induced, completely, a low weight w_t for the particle. After permutation (bottom row), the first particle is very close to the real state (and consequentially has high weight) and the second is far enough from it (and thus has a very low weight). After resampling, the latter will probably be deleted and only the first particle will remain, duplicated, close to the mode of the distribution $p(\mathbf{x}_{0:t}|\mathbf{y}_{1:t})$.

It is shown in [DUB 12b] that executing the permutation operation described above does not change the distribution. Given that the $\mathcal{X}^m$ are continuums, any two sub-particles have the same $\mathbf{Pa}(\mathbf{x}^m_t)$ value if and only if they have been generated by resampling from the same particle. In this case, they also have the same $\mathbf{Pa}(\mathbf{x}^m_{0:t})$ value. We allow ourselves to swap particles if they have the same $\mathbf{Pa}(\mathbf{x}^m_t)$ value. The advantage here is that we no longer have to consider the trajectories of the particles from time 0 to t, but only over the time interval t.

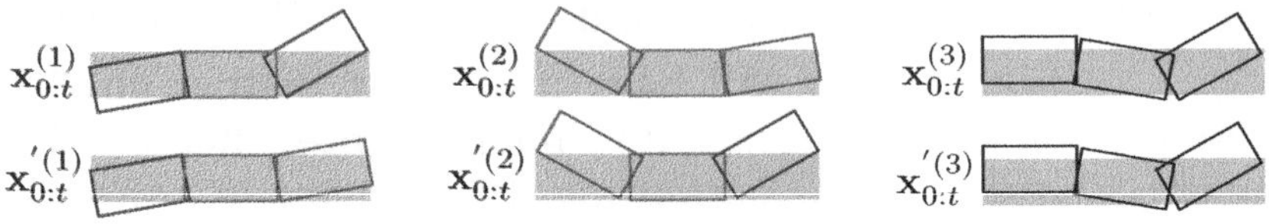

Figure 4.4. *Illustration of the permutation operation. Top row: three particles before permutation, only the two first of which can be swapped, as they have the same central particle (in green). The gray particles represent the real object. Note that the right part of the particle* $\mathbf{x}_{0:t}^{(1)}$ *is erroneous, while its left part is correct. Hence, the right parts of* $\mathbf{x}_{0:t}^{(1)}$ *and* $\mathbf{x}_{0:t}^{(2)}$ *are swapped to obtain new particles (bottom row): a very correct one* $\mathbf{x}_{0:t}^{\prime(1)}$ *and the another one less so* $\mathbf{x}_{0:t}^{\prime(2)}$. *For a color version of the figure, see www.iste.co.uk/dubuisson/tracking.zip*

4.3.2. *Decrease the number of resamplings*

Once we apply the basic algorithm of PS, we consider every part independently; we propagate, correct and resample this part, as well as those that have already been processed. The problem is that by multiplying the processing sub-spaces, we also multiply the number of resampling stages, which introduces a lot of noise into the estimation process and degrades performance over time. The idea, here, initiated in [GON 11] and completely finalized in [DUB 12b], is to simultaneously propagate/correct all the parts that are independent, and then execute a single resampling operation for all of these parts. By taking the example of Figure 4.3, we will process the torso (in green), then simultaneously the head and the two hands (in red) and finally the two forearms (in blue). In total, $K = 3$ resamplings will be executed instead of $M = 6$.

4.3.3. *General algorithm and results*

The general diagram of the method suggested in [DUB 14] is described in Figure 4.5: we kept the same color code as in

Figure 4.3 for more readability. The sign $\rightleftharpoons$ symbolizes the permutation operation and the sign $\sim$ the resampling one. Note that contrary to what was done in Algorithm 4.1, in which parts were processed one after the other, in our case, we simultaneously process partitions, i.e. subsets of parts. We need to introduce the following notations to generalize the PS algorithm (we omit here time indices as we assume that the defined sets do not change over time):

– P_k the partition k, $k = 1, \ldots, K$, thus the set of parts considered at the current instant;

– $Q_k = \sum_{h=1}^{k} P_h$ the set containing partitions that have already been processed and the parts of the current partition;

– $R_k = \sum_{h=k+1}^{K} P_h$ th set of partitions that remain to be processed.

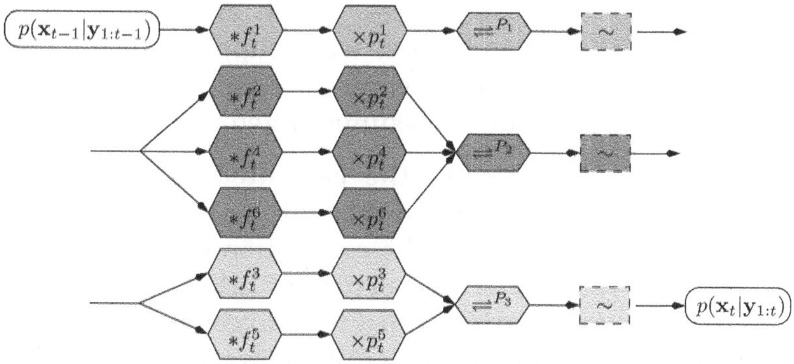

Figure 4.5. *Processing scheme for the example of Figure 4.3: the independent parts are propagated ($*f_t^i$) and corrected ($\times p_t^i$) in parallel, then swapped between themselves ($\rightleftharpoons$) to compose the best particles and finally resampled ($\sim$)*

Hence, we denote $\mathbf{x}_t^{P_k}$ the sub-states corresponding to the parts contained in P_k at instant t. By taking the example in Figure 4.3, we have $K = 3$ partitions, with $P_1 = \{1\}$ (in green), $P_2 = \{2, 4, 6\}$ (in red) and $P_3 = \{3, 5\}$ (in blue). Let us assume that we are about to process partition 2 at instant t; in this case, we have $P_2 = \{2, 4, 6\}$, $Q_2 = P_1 \cup P_2 = \{1, 2, 4, 6\}$, $R_2 = P_3 = \{3, 5\}$ and $\mathbf{x}_t^{P_2} = \{\mathbf{x}_t^2, \mathbf{x}_t^4, \mathbf{x}_t^6\}$. This leads to

Algorithm 4.3, which consists of applying the PS algorithm over these partitions, by adding a permutation stage.

The results in terms of computation time and quality of estimation (i.e. of tracking) of this algorithm have been compared to those of classical PS, as well as to the annealed PF (APF). The layers of simulated annealing of the APF were also replaced by a permutation stage to compare directly an optimization approach (looking for likelihood modes – see brief description in Chapter 3) to this state-space decomposition approach. The complete results obtained over synthetic and real sequences are given in [DUB 12b]. The synthetic sequences (examples of which can be found in Figure 4.6) allowed to test the robustness of the approach, as we were able to generate articulated objects whose number of arms, as well as their length, were parametric. Tests were also made over video sequences, which enabled us to test under real tracking conditions, with supplementary image constraints. In particular, several interesting properties of this algorithm were shown, as summarized below:

– it makes it possible for the permuted sample to focus better on the density modes to estimate (see examples given in Figure 4.9), in particular, the estimation errors decreased. This decrease relative to the errors of other approaches is greater when the object is defined in a larger space, therefore showing that the approach is better adapted to these cases. Figure 4.9 gives the tracking results obtained with $N = 2{,}000$ particles over a real video sequence containing two articulated objects, where we can see that this approach (right column) is more efficient for tracking than PS or APF (with two layers of simulated annealing). For example, it allows to reduce the resampling time by 60% and the total tracking times over the entire sequence by 43% in comparison to those of APF. Moreover, the estimation error is reduced by 18% in relation to APF with two layers of simulated annealing and 35% when compared to the results from PS;

Algorithm 4.3: Algorithm of swapping-based partitioned sampling (SBPS) [DUB 12b]

Input : set of particles $\{\mathbf{x}_{t-1}^{(i)}, w_{t-1}^{(i)}\}_{i=1}^{N}$, partition $\{P_1, \dots, P_K\}$
Output : set of particles $\{\mathbf{x}_{t}^{(i)}, w_{t}^{(i)}\}_{i=1}^{N}$

$\left\{\mathbf{x}_{t-1}^{(i)}, 1/N\right\}_{i=1}^{N} \leftarrow$ multinomial resampling of $\left\{\mathbf{x}_{t-1}^{(i)}, w_{t-1}^{(i)}\right\}_{i=1}^{N}$

for $k = 1$ **to** K **do**
  for each $m \in P_k$ **do**
    for $i = 1$ **to** N **do**
      Propagation: $\mathbf{x}_t^{m,(i)} \sim q\left(\mathbf{x}_t^m | \langle \mathbf{x}_{t-1}^{m,(i)}, \mathsf{Pa}(\mathbf{x}_t^{m,(i)}) \rangle, \mathbf{y}_{1:t}\right)$
    for each *subset* S_t *of particles belonging to* $\left\{\mathbf{x}_t^{m,(i)}\right\}_{i=1}^{N}$ *having the same* $\boldsymbol{Pa}(\mathbf{x}_t^{m,(i)})$ *values* **do**
      $\left\{\langle \mathbf{x}_t^{m,(i)}, \mathsf{Desc}(\mathbf{x}_{t-1}^{m,(i)}) \rangle, w_t^{m,(i)}\right\} \leftarrow$ permutation
      $\left\{\langle \mathbf{x}_t^{m,(i)}, \mathsf{Desc}(\mathbf{x}_{t-1}^{m,(i)}) \rangle, w_t^{m,(i)} : i \in S_t\right\}$
  $\left\{\langle \mathbf{x}_t^{Q_k,(i)}, \mathbf{x}_{t-1}^{R_k,(i)} \rangle, w_t^{(i)}\}_{i=1}^{N}\right\} \leftarrow$ weighted resampling of $\left\{\langle \mathbf{x}_t^{Q_k,(i)}, \mathbf{x}_{t-1}^{R_k,(i)} \rangle, w_t^{(i)}, \rho_t^{(i)}\right\}_{i=1}^{N}$
  Normalization: $\tilde{w}_t^{(i)} = \frac{w_t^{(i)}}{\sum_{j=1}^{N} w_t^{(j)}}$

Correction:

$$w_t^{(i)} \propto \tilde{w}_t^{(i)} \frac{\prod_{m=1}^{M} p\left(\mathbf{y}_t^m | \mathbf{x}_t^{m,(i)}\right) p\left(\mathbf{x}_t^{m,(i)} | \langle \mathbf{x}_{t-1}^{m,(i)}, \mathsf{Pa}(\mathbf{x}_t^{m,(i)}) \rangle\right)}{q\left(\mathbf{x}_t^{m,(i)} | \langle \mathbf{x}_{t-1}^{m,(i)}, \mathsf{Pa}(\mathbf{x}_t^{m,(i)}), \mathbf{y}_{1:t} \rangle\right)} \quad i = 1, \dots, N$$

Normalization: $\left\{w_t^{(i)}\right\}_{i=1}^{N} \leftarrow \left\{\frac{w_t^{(i)}}{\sum_{j=1}^{N} w_t^{(j)}}\right\}_{i=1}^{N}$

– it reduces the computation times, a phenomenon that becomes more notable 0permutation stage does not increase the computation time, as it is largely compensated by the time gain due to the decrease in the number of resamplings. Figures 4.7(a) and (b) show the computation time curves obtained over a synthetic sequence, according to the dimension of the state space $|\mathcal{X}| = 4K + 1$ (K being the data in abscissa). We can see that the greater the dimension, the more important the computation time regained with this approach.

Figure 4.7(c) gives the tracking errors (in pixels). Here again, we can see that the smallest errors are obtained with this approach;

– it makes it possible to follow several articulated objects simultaneously without having to multiply the number of filters, which is normally done to avoid having the filter manage a state space that is too large. The fact that this filter can incorporate all the objects within the same architecture shows its capacity to manage large state spaces;

– it is shown in [DUB 12b] that the estimated distribution is not modified by the permutation operation. This is very important, as a good number of algorithms modify this distribution, and thus exit the strict theoretical framework of the PF.

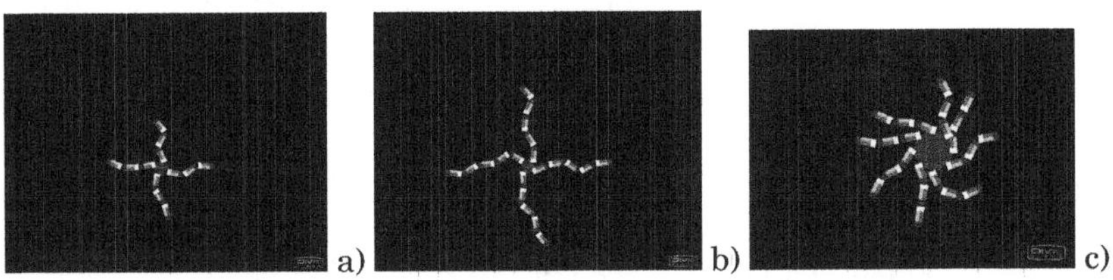

Figure 4.6. *Three close-ups on images extracted from the used synthetic sequences. The objects deform and move over time. The length of the arms or the number of parts that we process simultaneously changes over time, giving different dimensions of state space: a) $|\mathcal{X}| = 39$, b) $|\mathcal{X}| = 63$ and c) $|\mathcal{X}| = 75$. For a color version of the figure, see www.iste.co.uk/dubuisson/tracking.zip*

This sub-sample permutation technique was exploited to suggest a new resampling algorithm, which we explain in the following section.

4.4. Combinatorial resampling

A new resampling method, called combinatorial resampling and suggested in [DUB 12c], exploits the characteristics of articulated objects to resample over a sample of potentially exponential size, created implicitly (i.e.

we sample along this set without constructing it), which represents better the density to estimate. Based on the permutation idea described in section 4.3.1, this sample contains particles issued from all the possible permutations.

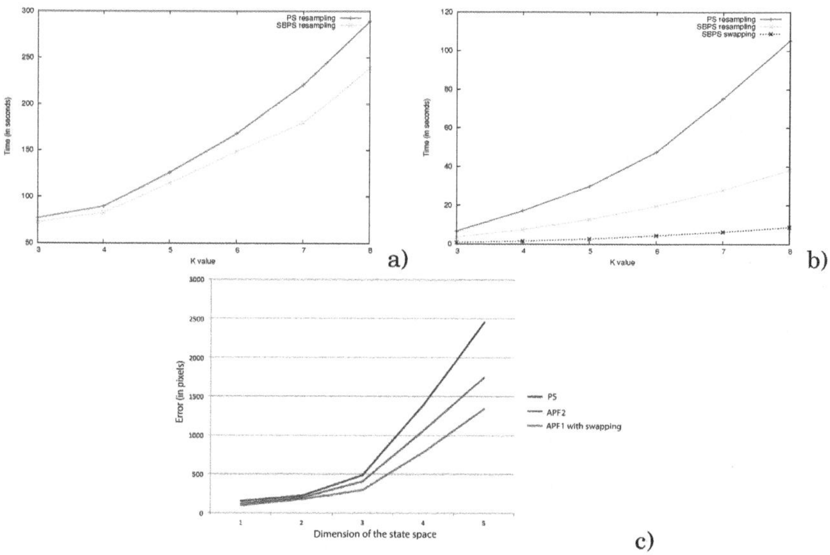

Figure 4.7. *Comparison of the results tracked in the synthetic sequences with $N = 600$, according to the dimension of the state space. a–b) On the left, the complete processing times over the entirety of the sequence (PS in red, approach suggested in [DUB 14], SBPS, in green), on the right, the times of resampling (PS in red, approach suggested in [DUB 14], SBPS, in green) and of permutation, which only concerns our approach (in blue). c) Tracking errors (in pixels) obtained by PS (in blue), APF with two layers (in green) and APF with integrated permutation (in red). For a color version of the figure, see www.iste.co.uk / dubuisson / tracking.zip*

In the example of Figure 4.3, let $\mathbf{x}_t^{(1)} = \langle 1, 2, 3, 4, 5, 6 \rangle$, $\mathbf{x}_t^{(2)} = \langle 1, 2', 3', 4', 5', 6' \rangle$ and $\mathbf{x}_t^{(3)} = \langle 1'', 2'', 3'', 4'', 5'', 6'' \rangle$ be three particles, where the numbers $1, 1', 1'', 2, 2', 2''$, etc., represent the state of a sub-part. Assume that we are at the second part of the PS algorithm with permutation, i.e. such that the set of currently processed parts of the object is $P_2 = \{2,4,6\}$ and we have the following sample $S = \{\mathbf{x}_t^{(1)}, \mathbf{x}_t^{(2)}, \mathbf{x}_t^{(3)}\}$:

$$\langle 1,2,3,4,5,6\rangle,\ \langle 1,2',3',4',5',6'\rangle,\ \langle 1'',2'',3'',4'',5'',6''\rangle$$

Parts $\{2,3\}$, $\{4,5\}$ and $\{6\}$ can be swapped in $\mathbf{x}_t^{(1)}$ and $\mathbf{x}_t^{(2)}$ as their torso, i.e. 1, is identical and we define S' as the union of all these permutation over S, which corresponds to the sequence described in Figure 4.8.

In practice, it is impossible to construct S', as the number of possible permutations increases exponentially with N. We therefore suggest, and this is the major idea of this contribution, to sample over S' without having to construct it and to this end, we will just calculate the weights of the identical central parts. To explain this, let us take the example illustrated in Figure 4.10, corresponding to the case of the object in Figure 4.3, part 6 of which we will not consider for simplification and use the notations defined in section 4.3.3. Assume that we are at stage k and that parts $P_k = \{3,5\}$ (i.e. the two forearms) have just been propagated and corrected (the parts that have already been resampled at the previous stage are, in this case, $Q_{k-1} = \{1,2,4\}$ and correspond to the central part mentioned above) and that we would like to resample over S', induced by S. To this end, we determine that first, the partition $S_1, \ldots, S_R$ of S, such that all the particles of S_r, $r = 1, \ldots, R$ have the same values of $\mathbf{x}_t^{Q_{k-1}}$. Hence, for example, in Figure 4.10, S_1 contains the two first particles, as their $\mathbf{x}_t^{Q_{k-1}}$ values are both $\langle 1,0,3\rangle$. A weight W_r (see its definition further on) is associated with every set S_r such that the sampling of the particles in these sets according to the weights W_r is equivalent to a sampling of particles S according to their weight: we thus obtain a set of particles that estimates the same distribution as S. We then define, for all $r \in \{1, \ldots, R\}$, the set S_r^m of particles of S whose part m is compatible with the value $\mathbf{x}_t^{Q_{k-1}}$ in S_r. S_r^m is therefore the set of particles S that have the same $\mathbf{Pa}(\mathrm{x}_t^m)$ value as those of S_r. For example, in Figure 4.10, S_1^3 is the set of the three first particles, as all of them have 1 as value of part $m = 2$.

$$\begin{array}{lll}
\langle 1,2\ ,3\ ,4\ ,5\ ,6\ \rangle & \langle 1,2',3',4',5',6'\rangle & \langle 1'',2'',3'',4'',5'',6''\rangle \\
\langle 1,2\ ,3\ ,4\ ,5\ ,6'\rangle & \langle 1,2',3',4',5',6\ \rangle & \langle 1'',2'',3'',4'',5'',6''\rangle \\
\langle 1,2\ ,3\ ,4',5',6\ \rangle & \langle 1,2',3',4\ ,5\ ,6'\rangle & \langle 1'',2'',3'',4'',5'',6''\rangle \\
\langle 1,2\ ,3\ ,4',5',6'\rangle & \langle 1,2',3',4\ ,5\ ,6\ \rangle & \langle 1'',2'',3'',4'',5'',6''\rangle \\
\langle 1,2',3',4\ ,5\ ,6\ \rangle & \langle 1,2\ ,3\ ,4',5',6'\rangle & \langle 1'',2'',3'',4'',5'',6''\rangle \\
\langle 1,2',3',4\ ,5\ ,6'\rangle & \langle 1,2\ ,3\ ,4',5',6\ \rangle & \langle 1'',2'',3'',4'',5'',6''\rangle \\
\langle 1,2',3',4',5',6\ \rangle & \langle 1,2\ ,3\ ,4\ ,5\ ,6'\rangle & \langle 1'',2'',3'',4'',5'',6''\rangle \\
\langle 1,2',3',4',5',6'\rangle & \langle 1,2\ ,3\ ,4\ ,5\ ,6\ \rangle & \langle 1'',2'',3'',4'',5'',6''\rangle
\end{array}$$

Figure 4.8. *Set of possible permutations for the case of three particles. For a color version of the figure, see www.iste.co.uk/dubuisson/tracking.zip*

The weights W_r correspond to the sum of the weights of particles that have already been generated by permutation in S_r. We define by $N_1, \ldots, N_R$ and $N_1^m, \ldots, N_R^m$ the respective sizes of $S_1, \ldots, S_R$ and $S_1^m, \ldots, S_R^m$. Finally, let $N^m = \max\{N_1^m, \ldots, N_R^m\}$ and, for all $r \in \{1, \ldots, R\}$, W_r^m the sum of weights assigned to the part m of particles S_r^m, i.e. $W_r^m = \sum_{\mathbf{x}_t^{(i)} \in S_r^m} w^{m,(i)}$. Then ,we can show [DUB 12c], for every r, that:

$$W_r = N_r \times \prod_{m \in P_k} \frac{N^m!}{A_{N_r^m}^{N_r}} \times A_{N_r^m-1}^{N_r-1} \times W_r^m \qquad [4.8]$$

where $A_n^k = n!/(n-k)!$ is the number of arrangements of k elements within n. Here, the expression is decomposed into two parts. The first, $\frac{N^m!}{A_{N_r^m}^{N_r}}$, represents the number of times where S_r needs to be duplicated in order for S' to estimate the same law as S (see column on the right in Figure 4.8). The second corresponds to the sum of the weights of the particles resulting from all the permutations of S_r and is equal to $A_{N_r^m-1}^{N_r-1} \times W_r^m \times N_R$.

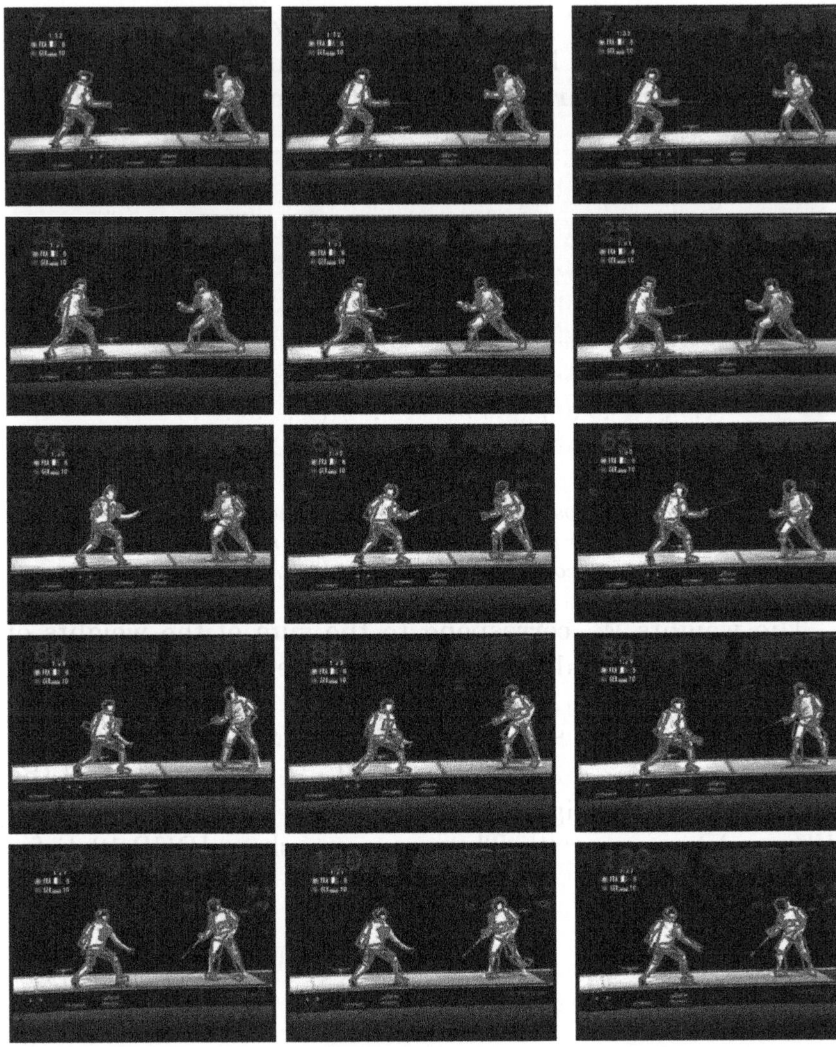

PS APF (two layers) PS with permutation

Figure 4.9. *Tracking results (close-ups) obtained by three approaches over the sequence* `Fencing` *of base UCF50*[3]*, with N =1,000: from top to bottom, images 7, 25, 65, 80 and 120. The estimated articulated objects are represented in red and correspond to the weighted average of the particles. For a color version of the figure, see www.iste.co.uk / dubuisson / tracking.zip*

3 http://www.cs.ucf.edu/vision/public_html/data.html.

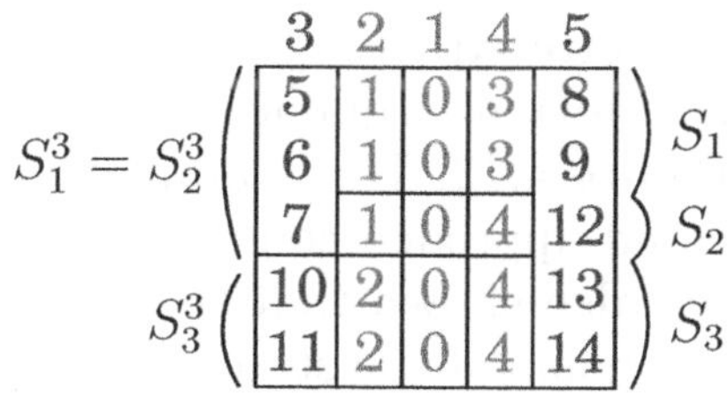

Figure 4.10. *The sets S_r and S_r^m. Every line represents a particle $\mathbf{x}_t^{(i)}$ and every column corresponds to a part m (sub-state $\mathbf{x}_t^{m,(i)}$) and contains the associated states (every number symbolizes a different value of $\mathbf{x}_t^m$). For a color version of the figure, see www.iste.co.uk / dubuisson / tracking.zip*

Then, it is enough, as we have shown, to resample according to the W_r. Algorithm 4.4 summarizes that the resampling procedure over S' ($\mathcal{M}(1; W_1, \ldots, W_R)$ designates a multinomial selection according to weights $\{W_r\}_{r=1}^R$). Note that if N is large, it is preferable to compute $\log(W_r)$ rather than W_r and the weights will then be $\exp(\log W_r - \log W)$, where $W = \max\{W_1, \ldots, W_R\}$.

Algorithm 4.4: Principe of combinatorial resampling [DUB 12c]

Input : set of particles $\{(\mathbf{x}_t^{Q_k,(i)}, \mathbf{x}_{t-1}^{R_k,(i)}, w_t^{(i)}\}_{i=1}^N$

Output : set of particles $\{(\mathbf{x}_t^{\prime\prime Q_k,(i)}, \mathbf{x}_{t-1}^{\prime\prime R_k,(i)}), w_t^{\prime\prime(i)}\}_{i=1}^N$

for $i = 1$ **to** N **do**

- $r \sim \mathcal{M}(1; W_1, \ldots, W_R)$
- $\mathbf{x}_t^{\prime\prime Q_{k-1},(i)} \leftarrow \mathbf{x}_t^{Q_{k-1},(z)}$ where $\mathbf{x}_t^{(z)} \in S_r$
- $w_t^{\prime\prime(i)} \leftarrow 1$
- **for each** m **in** P_k **do**
  - $j \sim \mathcal{M}\left(1; \{w_t^{m,(k)} : \mathbf{x}_t^{(k)} \in S_r^m\}\right)$
  - $\mathbf{x}_t^{\prime\prime m,(i)} \leftarrow \mathbf{x}_t^{m,(j)}$; $w_t^{\prime\prime(i)} \leftarrow w_t^{\prime\prime(i)} \times w_t^{m,(j)}$
  - $\mathbf{x}_{t-1}^{\prime\prime \mathbf{Desc}_{t-1}(\mathbf{x}_{t-1}^m),(i)} \leftarrow \mathbf{x}_{t-1}^{\mathbf{Desc}_{t-1}(\mathbf{x}_{t-1}^m),(j)}$

The resampling method was compared to the five described in Chapter 1, with real and synthetic sequences [DUB 12c]; better tracking stability was observed throughout all of the tested sequences, as shown by the tracking results in Figure 4.12. The examples in Figure 4.11 show that the suggested approach converges faster than the others. The tests have also shown that the approach is more efficient when the dimension of the state space is large and, *a fortiori*, when the number of particles necessary for tracking is high.

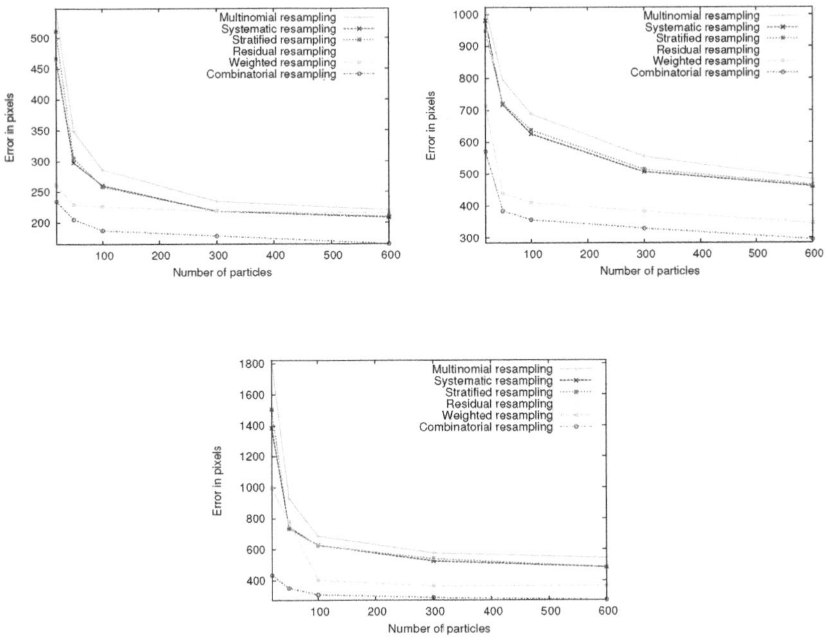

Figure 4.11. *Comparison of convergence results (i.e. as a function of N) over synthetic sequences according to the resampling method used. From left to right, for the object in Figure 4.6(a), the one in Figure 4.6(b) and the one in Figure 4.6(c). For a color version of the figure, see www.iste.co.uk/dubuisson/tracking.zip*

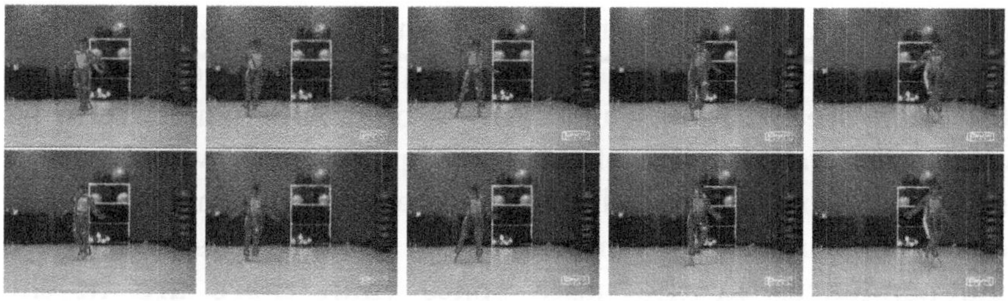

Figure 4.12. *Comparative tracking results obtained over the sequence* `JumpRope` *of base UCF50*[4]*, with* $N = 500$ *particles. Top row, by using residual resampling, and bottom row, by using this approach, combinatorial resampling. The estimated articulated object is represented in red and corresponds to the weighted average of the particles. For a color version of the figure, see www.iste.co.uk / dubuisson / tracking.zip*

4.5. Conclusion

In this chapter, we suggested three approaches exploiting the decomposition of the state space, all of which rely on the scheme based on PS. A first approach [WID 10b, WID 12], dedicated to multi-object tracking, estimates conjointly the joint state of the objects and the order in which the objects need to be processed. This approach is shown to be competitive, even better than certain state of the art methods, such as PS, BPS and DPS. However, even if this has not occurred in practice, the addition of a scenario order vector into the state vector very likely limits the possibilities of the approach to the number of objects tracked (we have tested up to eight objects). The two others, used in the context of tracking one or several articulated objects, rely on a principle of simultaneous processing and the permutation of sub-samples defined in the independent sub-spaces and proving a first solution for the estimation [DUB 12b, DUB 11b, GON 11] and a second one for resampling [DUB 12c]. This idea of permutation makes it possible to, in particular, overcome a first barrier by

4 http://www.cs.ucf.edu/vision/public_html/data.html.

estimating densities better than the other state-of-the-art approaches, while reducing the computation time. Moreover, the suggested approach is shown to be more efficient than its competitors as the size of the state space grows, which shows that it is particularly adapted to these cases. The suggested model is completely general as it does not assume more independences in the state space than there are in the considered problem. Additionally, its validity has been proven mathematically. There is still a more detailed study to do on its convergence: if we are capable of showing that the approach converges, it would be interesting to prove that it converges faster than PS, as we have shown in the tests made over synthetic and real sequences. Finally, it would also be interesting to show the capacities of the algorithm to focus on the distribution modes to estimate. This demonstration could be made using known multimodal densities and by measuring the Kullback–Leibler distance [KUL 51] between the estimated sample and the known sample. Indeed, criteria such as the average or the variance, commonly used for such demonstrations, are not adequate for measuring the capacity to sample a multimodal density.

5

Research Perspectives in Tracking and Managing Large Spaces

In this book, we have outlined a certain number of research works, generally from the past 15 years, concerned with tracking in video sequences by using sequential Monte Carlo methods. Today, many models using this stochastic formalism were suggested and are recognized and widely used, each addressing an important problem inherent to tracking. If today, their advantages are undeniable, the general consensus is that they are not, inherently, adapted to large state and observation spaces, which render more complex, or sometimes impossible, the computations to make. The guideline of the research presented in this book involves the management of large state and observation spaces. We suggested several contributions, described in Chapters 2, 3 and 4, providing solutions to data modeling and representation, as well as reducing the number of computations necessary in order to make it possible to integrate state vectors containing numerous parameters.

As we previously mentioned, there are robust and established tracking algorithms, some of which are integrated into commercial solutions for specific purposes. For instance, nowadays, traffic assessment on certain motorways is fully automatized: it is estimated via tracking algorithms. Companies commercialize surveillance systems, while video game consoles integrate cameras to augment the interactivity between the players and game. We could, therefore, wonder whether, strictly speaking, there is any more progress to make in tracking or not. Naturally, it will always be necessary to provide algorithms that are faster, more precise and more robust, however, in our opinion, the issue lies elsewhere. Tracking methods capable of managing large state spaces should position themselves in a more ambitious framework than simply being content to “just” track.

Today, video-surveillance is often associated with espionage and the loss of individual freedoms as described at the beginning of the previous century by George Orwell, in his novel 1984. Yet, there are many situations in which tracking could be extremely useful, especially if the goal is not to exploit it for identification, but rather for the purposes of analysis, in the specific context of individual behavior and/or relations between individuals and their environment. This analysis can take several forms. First of all, we can attempt to detect individual abnormal behavior in order to, for example, enhance safety in public (leaving a suspicious package) or private (a person looses consciousness at home) places. Video-surveillance can also be used for therapeutic purposes, as it can be used to analyze behaviors. This analysis makes it possible to, for instance, detect physical (abnormal gait) or mental (atypical social behavior) pathologies, as well as to verify the validity of a behavior (make sure that a patient took medicine). Video-surveillance, which will become commonplace during the upcoming years, should make it possible to improve our day-to-day safety and comfort. It is, therefore, necessary to develop new video analysis algorithms that would meet these needs. We describe in what follows, in terms of what we believe are the aims for tracking in the future, three themes that seem to be essential to the context of video-surveillance, and that relate to the analysis of individual human behavior (section 5.1), the detection of events in a scene (section 5.2) and the analysis of social interactions (section 5.3).

5.1. Tracking for behavioral analysis: toward finer tracking of the "future" and the "now"

Currently, the analysis of a video sequence featuring individuals usually aims to detect a particular behavioral phenomenon (for example, a person falling, a particular attitude, etc.). To this end, in general, a tracking algorithm is setup to trigger an alert once the phenomenon is detected. However, even detecting the phenomenon quite early (suppose a few images in advance), the alert would nonetheless be sent out too late. Still, every phenomenon has its own triggering factors, which can come from the persons or their environment. The objective is, therefore, not to detect the phenomenon that is taking place, but rather anticipate that it will take place. This requires considerably more detailed tracking that is interested in modeling the onset of a phenomenon rather than the phenomenon itself; a considerably more arduous task, which assumes that there will be a lot of data to process (specifically, several view-points). Such thorough analysis is also necessary while the phenomenon takes place, in order to understand the manner in which it occurs. For instance, doctors working on Parkinson's disease need to understand what causes a fall and what the precursory symptoms are. In particular, if they are interested in distinguishing between an accidental loss of balance (in this case, the person will be conscious and exhibit a behavior limiting the shock received the fall, such as stretching out arms) and a loss of balance due to the disease (in this case, the fall will be the result of a loss of consciousness and the person exhibit has no reaction).

Today, the only works that make it possible to perform tracking with precision are those using data from the motion capture technique. Here, the trajectories of approximately 50 points on the body (in general, these are articulations, along with the face, hands, etc.) are extracted, and then analyzed. By comparing these data to those issued from learning, certain psychological or physiological information about the person is inferred, such as by weight (for example, by studying the spatial repartition of certain points), gender (by analyzing gait, which differs on average between males and females) or action types (person is running, turning, etc.) [LIV 12]. The first published works prove that movement tells a lot more than what we have used it until now, however, it requires an experimental protocol that is much more constraining. We would like to orient our research toward using tracking for this sort of application, but with less intrusive systems for movement capture, as these analyses would need to operate over sequences showing people in their daily lives. As we have explained, it seems to us that the challenge is not tracking to detect (for example, a fall), but tracking to prevent or describe. These problems are important to a number of communities, which do not have access to automatic processing that can be provided by computer vision, but require tracking that is finer and more precise than currently available. Tracking should, of course, take into account both the spatial configuration and the persons appearance, as well as external information, coming from entities or other surrounding people, as the movement or the attitude of a person, outside of serious pathological cases, is influenced by their surroundings. Inevitably, the wealth of information contained in representation leads to large state and observation spaces. The important question to address here relates to the way of using (and above all, integrating) this wealth within the framework of the particle filter.

It is, therefore, necessary to consider new ways of describing information within sequences, in particular, by using $3D$, or even temporal $3D$, descriptions, as suggested by two recent comparative studies, which are concerned with either facial expressions [SAV 12], or recognizing actions [RUD 12]. The rapidly growing development of approaches aiming to extract "sub-gestures" or actions seem to indicate that a part of the community is going this way [LIU 12, SAG 12, THI 12], and some of these action primitives were recently introduced into the particle filter [HUS 07]. The saliency maps are good descriptors that have not been used enough in tracking yet. Their downfall is that they are often obtained by learning, and are therefore dedicated to a specific applicative context (for example, characterizing colors [WEI 06] or the foreground [LUO 12]). Still, recent approaches attempt to move away from learning. For example, in [VIK 12], the saliency maps are evaluated through sampling the rectangular regions of the image. This technique seems to be easily adaptable to the framework of particle filtering: the weights of the particles, each designating a rectangular region, will be evaluated as a function of the salient characteristics and used to refine the tracking. Finally, working with biologically inspired characteristics could make it possible to better deconstruct movement, as this same deconstruction corresponds to the human cognitive process of recognition. Works in [ESC 12] have in particular shown the advantage of the characteristics inspired from biology for recognizing actions.

Naturally, by taking into account all of these informations, the problem becomes a lot more complex to solve. In particular, the densities that we aim to estimate are defined in very large, non-parametric and multimodal spaces. It is, therefore, necessary to pursue works on modeling and to lead multidisciplinary endogenous works with the aim of enhancing theoretical models by combining techniques from various disciplines issued from applied mathematics, such as artificial intelligence and operations research. Bayesian networks and their properties for tracking appear to allow for more flexible models and, above all, correct, simple and efficient algorithms. We think that it would also be very interesting to use optimization techniques to improve the distribution sampling. If their use in the context of particle filtering was introduced many years ago (see, in particular, the introduction in Chapter 3), one of them yielding the annealing particle filter [DEU 00], we think that there are other methods, less known in the domain of computer vision, that could be applied. Unfortunately, the problem with annealing and a large amount of other optimization methods introduced into the particle filter is that they only allow us to find a single distribution mode, which is not necessarily adapted to the case of large state and observation spaces, where the filtering density is multimodal. Therefore, it is necessary to use "intelligent" techniques for exploring the solution space, which would exploit the information supplied by the particle filter (in particular, the weight of the particles), to determine the regions to explore. For example, a "tabu" search, which forbids returning into an already explored region of the space, could be used and would allow us to pursue local search after having found a local minimum, while storing this information.

5.2. Tracking for event detection: toward a top-down model

We present in this section another research perspective that adopts an opposed methodological view and is part of the applicative framework of detecting events. Tracking algorithms are rarely applicable in the real world. For example, external conditions (weather, crowd density, noisy backgrounds, etc.) are not yet fully mastered and integrated into the algorithms, despite the great progress made throughout the past 10 years. Furthermore, the real-time capacities are still a dream, with the exception of algorithms constructed entirely in an *ad hoc* fashion to respond to a specific problem, and therefore the parameters have already been optimized for the task at hand. Finally, there are no methods that are robust enough to make it possible to track one or more objects over sequences of several hours, which is necessary for video-surveillance applications. The key idea would be to not consider objects individually, but rather the $3D$ scene in its entirety. Only reasoning about the scene globally should allow us to overcome application-specific optimizations. Of course, this involves a representation that will be described, again, in a large space, which returns to the research made up to this point.

We should not be content with tracking-specific regions in a scene, but rather perform a global analysis of this scene. To this end, the use of the surrounding context provides as much information for guiding or analyzing the tracking, as the objects themselves, and the interaction between the two should be exploited more. Indeed, it is no longer simply an "actor", any entity that is distinguished from its environment by its dynamic or visual appearance and its individual behavior, that we are interested in, but its behavior in relation to its environment and every surrounding entity. Adopting this somewhat unusual point of view implicitly leads to problems that are complex to solve. Indeed, the classical approaches normally use bottom-up reasoning, which in general leads more toward a synthesis. In our case, we begin with local detections and analyses that we attempt to combine in order to obtain a global vision. We clearly consider top-down reasoning, whose aim is to perform an analysis of the problem, starting from a global, or meta, vision that is then cut down according to criteria that are not necessarily known in advance, into "typical" regions (for example, a region that contains a spatial, temporal or even a relational characteristic different from the others.)

The first problem that this creates is that we cannot consider using specific learning or modeling techniques for analysis or recognition, as this would require having a basis that reflects nearly every event that can occur within any scene. This means that we would not necessarily be able to detect things by what they are, but rather by the way they differ, spatially or temporally, from the rest, without necessarily attempting to name this difference. In this respect, the spatial relations between the entities and scene, and in particular, their temporal evolution, need to be exploited more. This will require being capable of comparing spatial relations, which are difficult at the current time. Extending spatial relations to modeling and tracking a $3D$ scene by integrating new types of more complex fuzzy spatial relations such as *Between*, or even *Along*, is a promising perspective. Finally, it will probably be interesting to attempt the extraction of collective movement from the scene, which characterizes mass behavior. Currently, this is essentially accomplished via techniques that analyze and observe the optical flow [WU 12], grouping into dense trajectories [KIM 11], or according to global trends [KIM 12], which may still be a little expensive, even though they can easily be parallelized. The top-down vision that we mentioned earlier would perhaps be a way of partitioning a $3D$ scene in relation to the collective movements detected in it.

The second problem is that we can no longer use models represented in spaces of fixed size. Indeed, it is necessary to consider that the objects can appear, disappear, fuse, separate, stop and start again. It is, therefore, essential to find a model that can be abstracted through *a priori* strong hypotheses. Exploiting spatio-temporal information, in particular the recursive estimations of the objects on the scene via a particle filter, is essential. Rather than considering, as accomplished previously, the introduction of birth, death, fusion or separation parameters into the state vector, estimated at every instant, we think that we could use types of models other than those supplied by simple Markov models. For example, non-stationary dynamic Bayesian networks make use of Markov chains possible, whose structure evolves over time, and have recently been used in [WAN 11b] to detect events in $1D$ signals.

5.3. Tracking to measure social interactions

In this section, we present several ideas in the way that we consider using tracking in the context of social interaction analysis. One of the new objectives for tracking is to allow us to better understand the contents of video sequences in terms of new semantics. If, for example, during the 1990s, we were interested in classifying facial expressions according to six typical expressions (joy, sadness, anger, fear, disgust and surprise) and the neutral expression, eventually with a degree of the expression's intensity, today it is necessary to distinguish the note of depression in sadness, and more generally, what differentiates an expression from an emotion. In fact, this is the finer analysis that we mentioned in section 5.1, which should make it possible to access a higher level of understanding. The analysis of social interactions is becoming of interest to a number of computer vision researchers. For example, analyzing the reaction of a person to a gesture or an environment is of major interest to understand social exchanges. Previously, the only solution offered for studying social interactions (both the way people behave between themselves and the way they can exchange) consisted of reviewing hours of video sequences showing groups of people interacting with each other. Therefore, in the upcoming years, tracking should make it possible to characterize social behaviors, interactions as well as exchanges, as it is, by the way, already beginning to be used.

First of all, knowledge outside of this field needs to be used. To take as an example, Paul Eckman, a psychologist, suggested at the end of the 1970s a representation system called *Facial Actions Coding System* (FACS) [EKM 78], which defines every possible facial action that can be achieved through muscle movement, combining a subset of which forms a facial expression. This system was used by the computer vision community at the beginning of the 2000s to help with the automatic analysis of facial expressions in images or image sequences [BLA 97, ESS 97, TIA 01]. Without the FACS, research would probably have not progressed as quickly as it did in this domain. Hence, the exogenous multidisciplinary approach aims to exploit knowledge external to the domain of computer vision in order to improve, within a specific domain of application, both the tracking model and the way it is used. Here, we reference experts from other disciplines, such as medicine, psychoanalysis, biology and sociology. This type of collaboration offers perspectives on extremely interesting works on social interactions, which we would like to study over the next few years, and give us, in particular, access to essential real data. New recent research avenues already took advantage of this multidisciplinary approach: recognizing a person's approval or disapproval through behavioral analysis [BOU 13], or measuring the synchronicity of an exchange (that is find out to what extent the action of one person – a gesture or speech, for example – was triggered, or not, by the action of another) [DEL 12].

Social interaction involves phenomena that are both physical, psychological and physiological, and only a multimodal analysis would make it possible to give a good sense of the measurement of social interactions within a group of people. It is necessary to consider the emotions of every person, which can be reflected by their face, behavior (for example, harsh gestures in the case of anger), as well as speech and attitudes (postures). It is also necessary to consider contact between people, the relations between their gestures, etc., and their contact with the environment.

In this case, confidence maps could be constructed for emotions, postures and interactions, and make it possible to find out about the importance of the presence of one or the other (or even all of them) within a time interval, and thus suggest rules for measuring a degree of social interaction. The true advantage of this approach is that it is integrated into the context of recursive sequential estimation, which adds temporal continuity. However, this necessitates a reflection on the choice or the definition of the descriptors that define emotions, postures or interactions. Currently, the privileged approach is learning: the recognition process first extracts characteristics, often spatio-temporal, which are then classified according to a basis. Bins for words [SIV 09] and gestures [PON 11] made their appearance and provide simplified models that made it possible to progress far in recognizing an attitude or a posture, for example. The major problem with this type of approach is that it cannot characterize what it has not learned: a new phenomenon will not be taken into account as it deviates from all the known rules. This is particularly awkward when we are specifically looking to measure interactions based on the analysis of a behavior that we do not necessarily know. It could be interesting to see if it is possible to characterize interactions other than by simply trying to recognize them, in order to generalize the process of analysis. For example, a frequency analysis of the trajectory of hands or the movement of the head would certainly bring equally importance, as it would allow us to highlight the phenomena of repetition, or the salient points of these trajectories.

We have presented three avenues of research that we believe are important to explore in the years to come. All of these avenues aim to exploit the information contained in images, or event sound, in order to refine tracking, whose goal could hence change. In particular, the integration of high-level information into the particle filter, whether the information is structural (of the spatial relation type), behavioral (posture) or emotional (mood), is necessary for the better understanding of the scene. Generally, multidisciplinary approaches seem to be an important factor to make progress in computer vision research and to overcome our current abilities. Indeed, it is this association of various competences that makes the originality of a research work.

Bibliography

[ABE 06] ABED A.E., DUBUISSON S., BÉRÉZIAT D., "Comparison of statistical and shape-based approaches for non-rigid motion tracking with missing data using a particle filter", *Advanced Concepts for Intelligent Vision Systems*, pp. 185–196, 2006.

[ABE 07a] ABED A.E., DUBUISSON S., BÉRÉZIAT D., "Energetic particle filter for online multiple target tracking", *International Conference on Image Processing*, pp. 493–496, 2007.

[ABE 07b] ABED A.E., DUBUISSON S., BÉRÉZIAT D., "ENMIM: energetic normalized mutual information model for online multiple object tracking with unlearned motions", *Advanced Concept on Intelligent Vision System*, pp. 955–967, 2007.

[ABE 08] ABED A.E., Suivi multi-objets par filtrage particulaire dans un contexte de données incomplétes et manquantes, PhD thesis, University of Pierre and Marie Curie, 2008.

[ABE 12] ABED A.E., DUBUISSON S., BÉRÉZIAT D., "Spatio-temporal target-measure association using an adaptive geometrical approach", *Pattern Recognition Letters*, vol. 33, no. 6, pp. 765–774, 2012.

[ADA 06] ADAM A., RIVLIN E., SHIMSHONI I., "Robust fragments-based tracking using the integral histogram", *IEEE Conference on Computer Vision and Pattern Recognition*, pp. 798–805, 2006.

[AHM 12] AHMADI M., MOJALLALI H., IZADI-ZAMANABADI R., "State estimation of nonlinear stochastic systems using a novel meta-heuristic particle filter", *Swarm and Evolutionary Computation*, vol. 4, pp. 44–53, 2012.

[AMB 07] AMBERG B., BLAKE A., FITZGIBBON A., *et al.*, "Reconstructing high quality face-surfaces using model based stereo", *International Conference on Computer Vision*, pp. 1–8, 2007.

[ANG 08] ANGELOVA D., MIHAYLOVA L., "Extended object tracking using Monte-Carlo methods", *IEEE Transactions on Signal Processing*, vol. 56, no. 2, pp. 825–832, 2008.

[ARN 05a] ARNAUD E., MÉMIN E., "An efficient Rao-Blackwellized particle filter for object tracking", *IEEE International Conference on Image Processing*, pp. 426–429, 2005.

[ARN 05b] ARNAUD E., MÉMIN E., CERNUSCHI-FRIAS B., "Conditional filters for image sequence-based tracking-application to point tracking", *Image Processing*, vol. 14, no. 1, pp. 63–79, 2005.

[ARN 07] ARNAUD E., MÉMIN E., "Partial linear Gaussian models for tracking in image sequences using sequential Monte-Carlo methods", *International Journal on Computer Vision*, vol. 74, no. 1, pp. 75–102, 2007.

[AVE 09] AVENEL C., MÉMIN E., PÉREZ P., "Tracking closed curves with non-linear stochastic filters", *Conference on Scale Space and Variational Methods*, pp. 576–587, 2009.

[BAB 09] BABENKO B., YANG M.-H., BELONGIE S., "Visual tracking with online multiple instance learning", *IEEE Conference on Computer Vision and Pattern Recognition*, pp. 983–990, 2009.

[BAD 07] BADRINARAYANAN V., PÉREZ P., CLERC F.L., *et al.*, "Probabilistic color and adaptive multi-feature tracking with dynamically switched priority between cues", *International Conference on Computer Vision*, pp. 1–8, 2007.

[BAL 82] BALLARD D., BROWN C., *Computer Vision*, Prentice Hall Inc., 1982.

[BAN 08] BANDOUCH J., ENGSTLER F., BEETZ M., "Evaluation of Hierarchical sampling strategies in 3D human pose estimation", *British Machine Vision Conference*, pp. 925–934, 2008.

[BAR 88] BAR-SHALOM Y., FORTMAN T., *Tracking and Data Association*, Academic Press, 1988.

[BAR 94] BARRON J., FLEET D., BEAUCHEMIN S., "Performance of optical flow techniques", *International Journal of Computer Vision*, vol. 12, no. 1, pp. 43–77, 1994.

[BAR 95] BAR-SHALOM Y., LI X., *Multitarget Multisensor Tracking: Principles and Techniques*, YBS Publishing, 1995.

[BAR 09] BARDET F., CHATEAU T., RAMADASAN D., "Illumination aware MCMC particle filter for long-term outdoor multi-object simultaneous tracking and classification", *International Conference on Computer Vision*, pp. 1623–1630, 2009.

[BER 06] BERNIER O., CHEUNG-MON-CHANG P., "Real-time 3D articulated pose tracking using particle filtering and belief propagation on factor graphs", *British Machine Vision Conference*, vol. 1, pp. 27–36, 2006.

[BER 09] BERNIER O., CHEUNG-MON-CHAN P., BOUGUET A., "Fast nonparametric belief propagation for real-time stereo articulated body tracking", *Computer Vision and Image Understanding*, vol. 113, pp. 29–47, 2009.

[BER 11] BERCLAZ J., FLEURET F., TURETKEN E., *et al.*, "Multiple object tracking using k-shortest paths optimization", *IEEE Transactions on Pattern Analysis and Machine Intelligence*, vol. 33, no. 9, pp. 1806–1819, 2011.

[BER 12] BERGER R., DUBUISSON S., GONZALES C., "Fast multiple histogram computation using Kruskal's algorithm", *International Conference on Image Processing*, pp. 1–4, 2012.

[BES 09] BESADA-PORTAS E., PLIS S., CRUZ J., LANE T., "Parallel subspace sampling for particle filtering in dynamic Bayesian networks", *European Conference on Machine Learning and Principles and Practice of Knowledge Discovery in Databases*, pp. 131–146, 2009.

[BHA 43] BHATTACHARYYA A., "On a measure of divergence between two statistical populations defined by probability distributions", *Bulletin of the Calcutta Mathematical Society*, vol. 35, pp. 99–110, 1943.

[BHA 09] BHANDARKAR S.M., LUO X., "Integrated detection and tracking of multiple faces using particle filtering and optical flow-based elastic matching", *Computer Vision and Image Understanding*, vol. 113, no. 6, pp. 708–725, 2009.

[BIR 98] BIRCHFIELD S., "Elliptical head tracking using intensity gradients and color histograms", *International Conference on Computer Vision*, pp. 232–237, 1998.

[BIS 07] BISHOP C.M., *Pattern Recognition and Machine Learning (Information Science and Statistics)*, Springer, 2007.

[BLA 97] BLACK M., YACOOB Y., "Recognizing facial expressions in image sequences using local parameterized models of image motion", *International Journal on Computer Vision*, vol. 25, no. 1, pp. 23–48, 1997.

[BLA 98a] BLACK M., JEPSON A., "Recognizing temporal trajectories using the CONDENSATION algorithm", *International Conference on Automatic Face and Gesture Recognition*, pp. 16–21, 1998.

[BLA 98b] BLAKE A., ISARD M., *Active Contours*, Springer-Verlag, 1998.

[BLA 99a] BLAKE A., NORTH B., ISARD M., "Learning multi-class dynamics", *Conference on Advances in Neural Information*, pp. 389–395, 1999.

[BLA 99b] BLANZ V., VETTER T., "A morphable model for the synthesis of 3D faces", *International Conference on Computer Graphics*, pp. 187–194, 1999.

[BLA 99c] BLANZ V., VETTER T., "A morphable model for the synthesis of 3D faces", *Conference on Computer Graphics and Interactive Techniques*, ACM Press/Addison-Wesley Publishing Co., pp. 187–194, 1999.

[BLO 96] BLOCH I., "Combination operators for data fusion: a comparative review with classification", *IEEE Transactions on Systems, Man, and Cybernetics, Part A*, vol. 26, no. 1, pp. 52–67, 1996.

[BLO 05] BLOCH I., "Fuzzy spatial relationships for image processing and interpretation: a review", *Image and Vision Computing*, vol. 23, no. 2, pp. 89–110, 2005.

[BLO 06] BLOCH I., COLLIOT O., CESAR R., "On the ternary spatial relation between", *IEEE Transactions on Systems, Man, and Cybernetics, Part B*, vol. 36, no. 2, pp. 312–327, 2006.

[BOU 13] BOUSMALIS K., MEHU M., PANTIC M., "Towards the automatic detection of spontaneous agreement and disagreement based on nonverbal behavior: a survey of related cues, databases, and tools", *Image and Vision Computing*, 2013.

[BRA 05] BRANSON K., BELONGIE S., "Tracking multiple mouse contours (without too many samples)", *IEEE Conference on Computer Vision and Pattern Recognition*, pp. 1039–1046, 2005.

[BRA 07a] BRASNETT P., MIHAYLOVA L., BULL D., *et al.*, "Sequential Monte-Carlo tracking by fusing multiple cues in video sequences", *Image and Vision Computing*, vol. 25, no. 8, pp. 1217–1227, 2007.

[BRA 07b] BRAY M., KOLLER-MEIER E., SCHRAUDOLPH N., *et al.*, "Fast stochastic optimization for articulated structure tracking", *Image and Visions Computing*, vol. 25, no. 3, pp. 352–364, 2007.

[BRA 07c] BRAY M., KOLLERMEIER E., VANGOOL L., "Smart particle filtering for high-dimensional tracking", *Computer Vision and Image Understanding*, vol. 106, no. 1, pp. 116–129, 2007.

[BRU 07] BRUBAKER M., FLEET D., HERTZMANN A., "Physics-based person tracking using simplified lower-body dynamics", *IEEE Conference on Computer Vision and Pattern Recognition*, pp. 1–8, 2007.

[BRU 09] BRUBAKER M., FLEET D., HERTZMANN A., "Physics-based person tracking using the anthropomorphic walker", *International Journal of Computer Vision*, vol. 87, nos. 1–2, pp. 140–155, 2009.

[BUG 08] BUGEAU A., PÉREZ P., "Track and cut: simultaneous tracking and segmentation of multiple objects with graph cuts", *EURASIP Journal on Image and Video Processing*, vol. 2008, no. 3, pp. 1–15, 2008.

[CAI 08] CAI Z., LIN Z., "Fuzzy particle pilter used for tracking of leukocytes", *International Symposium on Intelligent Information Technology Application Workshops*, pp. 562–565, 2008.

[CAR 10] CARNEIRO G., NASCIMENTO J., "Multiple dynamic models for tracking the left ventricle of the heart from ultrasound data using particle filters and deep learning architectures", *IEEE Conference on Computer Vision and Pattern Recognition*, pp. 2815–2822, 2010.

[CHA 08] CHANG W., CHEN C., JIAN Y., "Visual tracking in high-dimensional state space by appearance-guided particle filtering", *IEEE Transactions on Image Processing*, vol. 17, no. 7, pp. 1154–1167, 2008.

[CHA 10] CHANG I.-C., LIN S.-Y., "3D human motion tracking based on a progressive particle filter", *Pattern Recognition*, vol. 43, no. 10, pp. 3621–3635, 2010.

[CHE 01] CHEN Y., RUI Y., HUANG T., "JPDAF-based HMM for real-time contour tracking", *IEEE Conference on Computer Vision and Pattern Recognition*, pp. 543–550, 2001.

[CHE 02] CHEN Y., HUANG T., RUI Y., "Parametric contour tracking using unscented Kalman filter", *International Conference on Image Processing*, pp. 613–616, 2002.

[CHE 03a] CHECKA N., WILSON K., RANGARAJAN V., *et al.*, "A probabilistic framework for multi-modal multi-person tracking", *Workshop on Multi-Object Tracking*, pp. 100–106, 2003.

[CHE 03b] CHEN Z., Bayesian Filtering: from Kalman filters to particle filters, and beyond, Report, McMaster University, 2003.

[CHE 04] CHEN Y., RUI Y., "Real-time speaker tracking using particle filter sensor fusion", *Proceedings of the IEEE*, vol. 92, no. 3, pp. 485–494, 2004.

[CHE 08] CHENOUARD N., CHAUMONT F.D., BLOCH I., *et al.*, "Improving 3D tracking in microscopy by joint estimation of kinetic and image models", *International Conference on Medical Image Computing and Computer Assisted Intervention – Workshop*, pp. 1–8, 2008.

[CHO 09] CHOCKALINGAM P., PRADEEP N., BIRCHFIELD S., "Adaptive fragments-based tracking of non-rigid objects using level sets", *International Conference Computer Vision*, pp. 1530–1537, 2009.

[COL 05] COLLINS R.T., LIU Y., LEORDEANU M., "Online selection of discriminative tracking features", *IEEE Transactions on Pattern Analysis and Machine Intelligence*, vol. 27, no. 10, pp. 1631–1643, 2005.

[COM 00] COMANICIU D., RAMESH V., MEER P., "Real-time tracking of non-rigid objects using mean-shift", *IEEE Conference on Computer Vision and Pattern Recognition*, pp. 142–149, 2000.

[COM 03] COMANICIU D., RAMESH V., MEER P., "Kernel-based object tracking", *IEEE Transaction on Pattern Analysis and Machine Intelligence*, vol. 25, pp. 564–577, 2003.

[COV 00] COVELL M., RAHINI A., HARVILLE M., *et al.*, "Articulated-pose estimation using brightness- and depth-constancy constraints", *IEEE International Conference on Computer Vision and Pattern Recognition*, pp. 438–445, 2000.

[COX 93] COX I., "A review of statistical data association techniques for motion correspondence", *International Journal of Computer Vision*, vol. 41, no. 1, pp. 53–65, 1993.

[CUI 07] CUI J., ZHA H., ZHAO H., *et al.*, "Laser-based detection and tracking of multiple people in crowds", *Computer Vision and Image Understanding*, vol. 106, nos. 2–3, pp. 300–312, 2007.

[CZY 07] CZYZ J., RISTIC B., MACQ B., "A particle filter for joint detection and tracking of color objects", *Image and Vision Computing*, vol. 25, no. 8, pp. 1271–1281, 2007.

[DAI 04] DAI X., HAGER G., "A particle filter without dynamics for robust 3D face tracking", *IEEE Conference on Computer Vision and Pattern Recognition*, pp. 70–78, 2004.

[DAR 08a] DARBY J., LI B., COSTEN N., "Behaviour based particle filtering for human articulated motion tracking", *IEEE International Conference on Pattern Recognition*, pp. 1–4, 2008.

[DAR 08b] DARONKOLAEI A.G., NAZARI V., MENHAJ M.B., *et al.*, "A Joint Probability Data Association Filter Algorithm for Multiple Robot Tracking Problems", in FRITZSCHE P., (ed.), *Tools in Artificial Intelligence*, InTech, pp. 1–24, 2008.

[DAS 11] DAS S., MAITY S., QU B.-Y., *et al.*, "Real-parameter evolutionary multimodal optimization: a survey of the state-of-the-art", *Swarm and Evolutionary Computation*, vol. 1, no. 2, pp. 71–88, 2011.

[DEL 12] DELAHERCHE E., CHETOUANI M., MAHDHAOUI A., *et al.*, "Interpersonal synchrony: a survey of evaluation methods across disciplines", *IEEE Transactions on Affective Computing*, vol. 3, no. 3, pp. 349–365, 2012.

[DEU 00] DEUTSCHER J., BLAKE A., REID I., "Articulated body motion capture by annealed particle filtering", *IEEE Conference on Computer Vision and Pattern Recognition*, pp. 126–133, 2000.

[DEU 05] DEUTSCHER J., REID I., "Articulated body motion capture by stochastic search", *International Journal of Computer Vision*, vol. 61, pp. 185–205, 2005.

[DOR 05] DORNAIKA F., DAVOINE F., "Simultaneous facial action tracking and expression recognition using a particle filter", *International Conference on Computer Vision*, pp. 1733–1738, 2005.

[DOR 10] DORE A., BEOLDO A., REGAZZONI C., "Multitarget tracking with a corner-based particle filter", *International Conference on Computer Vision, Workshop*, pp. 1251–1258, 2010.

[DOU 00a] DOUCET A., DE FREITAS N., MURPHY K., *et al.*, "Rao-Blackwellised particle filtering for dynamic Bayesian networks", *Conference on Uncertainty in Artificial Intelligence*, pp. 176–183, 2000.

[DOU 00b] DOUCET A., GODSILL S., ANDRIEU C., "On sequential Monte-Carlo sampling methods for Bayesian filtering", *Statistics and Computing*, vol. 10, no. 3, pp. 197–205, 2000.

[DOU 01] DOUCET A., DE FREITAS N., GORDON N., *Sequential Monte-Carlo Methods in Practice*, Springer-Verlag, New York, 2001.

[DOU 05] DOUC R., CAPPÉ O., MOULINES E., "Comparison of resampling schemes for particle filtering", *International Symposium on Image and Signal Processing and Analysis*, pp. 64–69, 2005.

[DU 05] DU W., PIATER J., "Tracking by cluster analysis of feature points using a mixture particle filter", *IEEE Conference on Advanced Video and Signal Based Surveillance*, pp. 165–170, 2005.

[DUB 80] DUBOIS D., PRADE H., *Fuzzy Sets and Systems: Theory and Applications*, Academic Press, Inc., 1980.

[DUB 06a] DUBUISSON S., "An adaptive clustering for multiple object tracking in sequences in and beyond the visible spectrum", *IEEE Conference on Computer Vision and Pattern Recognition, Workshop*, pp. 142–150, 2006.

[DUB 06b] DUBUISSON S., "Recursive clustering for multiple object tracking", *IEEE International Conference on Image Processing*, pp. 2805–2808, 2006.

[DUB 09] DUBUISSON S., FABRIZIO J., "Optimal recursive clustering of likelihood functions for multiple object tracking", *Pattern Recognition Letters*, vol. 30, no. 6, pp. 606–614, 2009.

[DUB 10a] DUBUISSON S., "The computation of the Bhattacharyya distance between histograms without histograms", *International Conference on Image Processing Theory Tools and Applications*, pp. 373–378, 2010.

[DUB 10b] DUBUISSON S., "Tree-structured temporal information for fast histogram computation", *International Conference on Computer Vision Theory and Applications*, pp. 1–10, 2010.

[DUB 11a] DUBUISSON S., "Tree-structured image difference for fast histogram and distance between histograms computation", *Pattern Recognition Letters*, vol. 32, no. 3, pp. 411–422, 2011.

[DUB 11b] DUBUISSON S., GONZALES C., NGUYEN X., "Swapping-based partitioned sampling for better complex density estimation: application to articulated object tracking", *International Conference on Scalable Uncertainty Management*, pp. 525–538, 2011.

[DUB 12a] DUBUISSON S., GONZALES C., "Min-space integral histogram", *European Conference on Computer Vision*, pp. 188–201, 2012.

[DUB 12b] DUBUISSON S., GONZALES C., "An optimized DBN-based mode-focussing particle filter", *International Conference on Computer Vision and Pattern Recognition*, pp. 1934–1939, 2012.

[DUB 12c] DUBUISSON S., GONZALES C., NGUYEN X., "DBN-based combinatorial resampling for articulated object tracking", *Conference on Uncertainty in Artificial Intelligence*, 2012.

[DUB 14] DUBUISSON S., GONZALES C., NGUYEN X., "Sub-sample swapping for sequential Monte Carlo approximation of high-dimensional densities", *International Journal on Approximate Reasoning*, vol. 54, no. 7, pp. 934–953, 2014.

[DUF 09] DUFFNER S., ODOBEZ J., RICCI E., "Dynamic partitioned sampling for tracking with discriminative features", *British Machine Vision Conference*, pp. 1–9, 2009.

[DUN 10] DUNNE P., MATUSZEWSKI B., "Choice of similarity measure, likelihood function and parameters for histogram based particle filter tracking in CCTV grey scale video", *Image and Vision Computing*, vol. 29, nos. 2–3, pp. 178–189, 2010.

[EKM 78] EKMAN P., FRIESEN W., *Facial Action Coding System: A Technique for the Measurement of Facial Movement*, Consulting Psychologists Press, 1978.

[ERC 07] ERCAN A., GAMAL A.E., GUIBAS L., "Object tracking in the presence of occlusions via a camera network", *International Conference on Information Processing in Sensor Networks*, pp. 509–518, 2007.

[ERD 12a] ERDEM E., DUBUISSON S., BLOCH I., "Fragment based tracking with adaptive cue integration", *Computer Vision and Image Understanding*, vol. 116, no. 7, pp. 827–841, 2012.

[ERD 12b] ERDEM E., DUBUISSON S., BLOCH I., "Visual tracking by fusing multiple cues with context-sensitive reliabilities", *Pattern Recognition*, vol. 45, no. 5, pp. 1948–1959, 2012.

[ESC 12] ESCOBAR M.J., KORNPROBST P., "Action recognition via bio-inspired features: the richness of center-surround interaction", *Computer Vision and Image Understanding*, vol. 116, pp. 593–605, 2012.

[ESS 97] ESSA I.A., PENTLAND A.P., "Coding, analysis, interpretation and recognition of facial expressions", *IEEE Transactions on Pattern Analysis and Machine Intelligence*, vol. 19, no. 7, pp. 757–763, 1997.

[FAG 08] FAGGIAN N., PAPLINSKI A., SHERRAH J., "3D Morphable model fitting from multiple views", *International Conference on Automatic Face and Gesture Recognition*, pp. 1–6, 2008.

[FAU 06] FAUX F., LUTHON F., "Robust face tracking using colour Dempster-Shafer fusion and particle filter", *International Conference on Information Fusion*, pp. 1–7, 2006.

[FEA 02] FEARNHEAD P., "MCMC, Sufficient statistics and particle filters", *Journal of Computational and Graphical Statistics*, vol. 11, no. 4, pp. 848–862, 2002.

[FLE 08] FLEURET F., BERCLAZ J., LENGAGNE R., *et al.*, "Multi-camera people tracking with a probabilistic occupancy map", *IEEE Transactions on Pattern Analysis and Machine Intelligence*, vol. 30, no. 2, pp. 267–282, 2008.

[FOR 83] FORTMANN T., BAR-SHALOM Y., SCHEFFE M., "Sonar tracking of multiple targets using joint probabilistic data association", *IEEE Journal of Oceanic Engineering*, vol. 8, no. 3, pp. 173–194, 1983.

[FRE 10] FREIFELD O., WEISS A., ZUFFI S., *et al.*, "Contour people: a parameterized model of 2D articulated human shape", *IEEE Conference on Computer Vision and Pattern Recognition*, pp. 639–646, 2010.

[GAL 06] GALL J., ROSENHAHN B., BROX T., *et al.*, Chapter Learning for multi-view 3D tracking in the context of particle filters, "*Advances in Visual Computing*", *LNCS*, vol. 4292, pp. 59–69, 2006.

[GAR 03] GARCIA J., MOLINA J., BESADA J., *et al.*, "Fuzzy approach for data association in image tracking", *Mathware and Soft Computing*, vol. 10, pp. 117–129, 2003.

[GAU 97] GAUVRIT H., CADRE J.-P.L., JAUFFRET C., "A formulation of multitarget tracking as an incomplete data problem", *IEEE Transactions on Aerospace and Electronic Systems*, vol. 33, pp. 1242–1257, 1997.

[GEL 04] GELGON M., BOUTHEMY P., CADRE J.L., "Recovery of the trajectories of multiple moving objects in an image sequence with a PMHT approach", *Journal of Image and Vision Computing*, vol. 23, no. 1, pp. 19–31, 2004.

[GID 08] GIDEL S., CHECCHIN P., BLANC C., *et al.*, "Pedestrian detection method using a multilayer laserscanner: application in urban environment", *International Conference on Intelligent Robots and Systems*, pp. 173–178, 2008.

[GIE 02] GIEBEL J., GAVRILA D., "Multimodal shape tracking with point distribution models", *DAGM Symposium on Pattern Recognition*, pp. 1–8, 2002.

[GON 07] GONCALVES W., MONTEIRO J., DE ANDRADE SILVA J., *et al.*, "Multiple mice tracking using a combination of particle filter and k-means", *Symposium on Computer Graphics and Image Processing*, pp. 173–178, 2007.

[GON 11] GONZALES C., DUBUISSON S., NGUYEN X., "Simultaneous partitioned sampling for articulated object tracking", *Advanced Concepts for Intelligent Vision Systems*, pp. 150–161, 2011.

[GOR 93] GORDON N., SALMOND D., SMITH A., "Novel approach to nonlinear/non-Gaussian Bayesian state estimation", *IEEE Proceedings of Radar and Signal Processing*, vol. 140, no. 2, pp. 107–113, 1993.

[GOY 10] GOYAT Y., CHATEAU T., TRASSOUDAINE L., "Tracking of vehicle trajectory by combining a camera and a laser rangefinder", *Machine Vision and Applications*, vol. 21, no. 3, pp. 275–286, 2010.

[GRA 05] GRAUMAN K., DARRELL T., "The pyramid matching kernel: discriminative classification with sets of image features", *International Conference on Computer Vision*, pp. 1458–1465, 2005.

[GRA 06a] GRABNER H., GRABNER M., BISCHOF H., "Real-time tracking via on-line boosting", *British Machine Vision Conference*, pp. 47–56, 2006.

[GRA 06b] GRANVILLE V., "Simulated annealing: a proof of convergence", *IEEE Transactions Pattern Analysis and Machine Intelligence*, vol. 16, no. 6, pp. 652–656, 2006.

[GRA 08] GRABNER H., LEISTNER C., BISCHOF H., "Semi-supervised on-line boosting for robust tracking", *European Conference on Computer Vision*, pp. 234–247, 2008.

[GUE 12] GUERRA-FILHO G., BISWAS A., "The human motion database: a cognitive and parametric sampling of human motion", *Image and Vision Computing*, vol. 30, pp. 251–261, 2012.

[HAF 95] HAFNER J., SAWHNEY H., EQUITZ W., *et al.*, "Efficient color histogram indexing for quadratic form distance function", *Pattern Analysis and Machine Intelligence*, vol. 17, no. 7, pp. 729–736, 1995.

[HAI 06] HAIBIN L., OKADA K., "Diffusion distance for histogram comparison", *IEEE Conference on Computer Vision and Pattern Recognition*, pp. 245–253, 2006.

[HAN 05a] HAN H., DING Y.-S., HAO K.-R., *et al.*, "An evolutionary particle filter with the immune genetic algorithm for intelligent video target tracking", *Computers and Mathematics with Application*, vol. 62, pp. 185–205, 2005.

[HAN 05b] HANSEN D., HAMMOUD R., SATRIA R., *et al.*, "Improved likelihood function in particle-based IR eye tracking", *IEEE Conference on Computer Vision and Pattern Recognition, Workshop*, pp. 5–12, 2005.

[HAU 10a] HAUBERG S., PEDERSEN K.S., "Stick it! Articulated tracking using spatial rigid object priors", *Asian Conference on Computer Vision*, pp. 758–769, 2010.

[HAU 10b] HAUBERG S., SOMMER S., PEDERSEN K., "Gaussian-like spatial priors for articulated tracking", *European Conference on Computer Vision*, pp. 425–437, 2010.

[HAU 11] HAUBERG S., PEDERSEN K.S., "Predicting articulated human motion from spatial processes", *International Journal of Computer Vision*, Springer, vol. 94, no. 3, pp. 317–334, 2011.

[HER 12] HEROLD C., DESPIEGEL V., DUBUISSON S., *et al.*, "Head shape estimation using a particle filter including unknown static parameters", *International Conference on Computer Vision Theory and Application*, pp. 284–293, 2012.

[HER 14] HEROLD C., DESPIEGEL V., GENTRIC S., *et al.*, "Recursive head reconstruction from multi-view video sequences", *Computer Vision and Image Understanding*, vol. 122, pp. 182–201, 2014.

[HOE 06] HOEY J., VON BERTOLDI A., POUPART P., *et al.*, "Tracking using flocks of features, with application to assisted handwashing", *British Machine Vision Conference*, vol. 1, pp. 367–376, 2006.

[HOE 10] HOEY J., POUPART P., BERTOLDI A., *et al.*, "Automated handwashing assistance for persons with dementia using video and a partially observable Markov decision process", *Computer Vision and Image Understanding*, vol. 114, no. 5, pp. 503–519, 2010.

[HOF 11] HOFMANN M., GAVRILA D., "3D Human model adaptation by frame selection and shape-texture optimization", *Computer Vision and Image Understanding*, vol. 115, no. 11, pp. 1559–1570, 2011.

[HU 08] HU Z., FAN X., SONG Y., *et al.*, "Joint trajectory tracking and recognition based on bi-directional nonlinear learning", *Image and Vision Computing*, vol. 27, no. 9, pp. 1302–1312, 2008.

[HUO 12] HUO F., HENDRIKS E., "Multiple people tracking and pose estimation with occlusion estimation", *Computer Vision and Image Understanding*, vol. 116, pp. 634–647, 2012.

[HUS 07] HUSZ Z., WALLACE A., "Evaluation of a hierarchical partitioned particle filter with action primitives", *IEEE Conference on Computer Vision and Pattern Recognition*, pp. 1–8, 2007.

[IHL 04] IHLER A.T., FISHER J.W., MOSES R.L., *et al.*, "Nonparametric belief propagation for self-calibration in sensor networks", *International Symposium on Information Processing in Sensor Networks*, pp. 225–233, 2004.

[ISA 98a] ISARD M., BLAKE A., "CONDENSATION – conditional density propagation for visual tracking", *International Journal of Computer Vision*, vol. 29, no. 1, pp. 5–28, 1998.

[ISA 98b] ISARD M., BLAKE A., "ICONDENSATION: unifying low-level and high-level tracking in a Stochastic framework", *European Conference on Computer Vision*, pp. 893–908, 1998.

[ISA 03] ISARD M., "PAMPAS: real-valued graphical models for computer vision", *IEEE Conference on Computer Vision and Pattern Recognition*, pp. 613–620, 2003.

[JAC 04] JACKSON J., YEZZI A., SOATTO S., "Tracking deformable moving objects under severe occlusions", *IEEE Conference on Decision and Control*, pp. 2990–2995, 2004.

[JAW 06] JAWARD M., MIHAYLOVA L., CANAGARAJAH N., *et al.*, "Multiple object tracking using particle filters", *IEEE Aerospace Conference*, pp. 8–12, 2006.

[JAZ 70] JAZWINSKI A. (ed.), *Stochastic Processes and Filtering Theory*, Academic Press, 1970.

[JOH 10] JOHN V., TRUCCO E., IVEKOVIC S., "Markerless human articulated tracking using hierarchical particle swarm optimization", *Image and Vision Computing*, vol. 28, no. 11, pp. 1530–1547, April 2010.

[JUL 97] JULIER S.J., UHLMANN J.K., "A new extension of the Kalman filter to nonlinear systems", *International Symposium on Aerospace/Defense Sensing, Simulation and Controls*, vol. 3, pp. 182–193, 1997.

[KAL 60] KALMAN R., "A new approach to linear filtering and prediction problems", *Transactions of the ASME – Journal of Basic Engineering*, vol. 82, pp. 35–45, 1960.

[KAM 05] KAMEL H., BADAWY W., "Fuzzy logic based particle filter for tracking a maneuverable target", *Midwest Symposium on Circuits and Systems*, vol. 2, pp. 1537–1540, 2005.

[KAN 95] KANAZAWA K., KOLLER D., RUSSELL S., "Stochastic simulation algorithms for dynamic probabilistic networks", *Conference on Uncertainty in Artificial Intelligence*, pp. 346–335, 1995.

[KAR 11] KARAVASILIS V., NIKOU C., LIKAS A., "Visual tracking using the Earth Mover's distance between Gaussian mixtures and Kalman filtering", *Image and Vision Computing*, vol. 9, no. 5, pp. 295–305, 2011.

[KAZ 09] KAZUHIRO H., "Adaptive weighting of local classifiers by particle filters for robust tracking", *Pattern Recognition*, vol. 42, pp. 619–628, 2009.

[KHA 06] KHAN Z., BALCH T., DELLAERT F., "MCMC data association and sparse factorization updating for real time multitarget tracking with merged and multiple measurements", *IEEE Transactions on Pattern Analysis And Machine Intelligence*, vol. 28, no. 12, pp. 1960–1972, 2006.

[KIM 07] KIM Y.-J., WON C.-H., PARK J.-M., *et al.*, "Fuzzy adaptive particle filter for a positioning beacon system", *International Conference on Control, Automation and Systems*, pp. 1198–1202, 2007.

[KIM 11] KIM K., LEE D., ESSA I., "Gaussian process regression flow for analysis of motion trajectories", *International Conference on Computer Vision*, pp. 1164–1171, 2011.

[KIM 12] KIM K., LEE D., ESSA I., "Detecting regions of interest in dynamic scenes with camera motions", *IEEE Conference on Computer Vision and Pattern Recognition*, pp. 1258–1265, 2012.

[KIT 96] KITAGAWA G., "Monte-Carlo filter and smoother for non-Gaussian nonlinear state space models", *Journal of Computational and Graphical Statistics*, vol. 5, no. 1, pp. 1–25, 1996.

[KJE 10] KJELLSTROM H., KRAGIC D., BLACK M., "Tracking people interacting with objects", *IEEE International Conference on Computer Vision and Pattern Recognition*, pp. 747–754, 2010.

[KNO 03] KNOPS Z., MAINTZ J., VIERGEVER M., *et al.*, "Normalized mutual information based registration using k-means clustering based histogram binning", *SPIE Medical Imaging*, vol. 5032, pp. 1072–1080, 2003.

[KON 03] KONSTANTINOVA P., UDVAREV A., SEMERDJIEV T., "A study of a target tracking algorithm using global nearest neighbor approach", *International Conference on Computer Systems and Technologies*, pp. 290–295, 2003.

[KRU 56] KRUSKAL J.B., "On the shortest spanning subtree of a graph and the traveling salesman problem", *Proceedings of the American Mathematical Society*, vol. 7, no. 1, pp. 48–50, 1956.

[KRZ 10] KRZESZOWSKI T., KWOLEK B., "Articulated body motion tracking by combined particle swarm optimization and particle filtering", *IEEE International Conference on Computer Vision and Pattern Recognition*, pp. 147–154, 2010.

[KUL 51] KULLBACK S., LEIBLER R., "On information and sufficiency", *Annals of Mathematical Statistics*, vol. 22, no. 1, pp. 79–86, 1951.

[KWO 08] KWON J., PARK F., "A geometric approach to particle filtering-based visual tracking", *International Conference on Ubiquitous Robots and Ambient Intelligence*, pp. 1–4, 2008.

[LAK 08] LAKAEMPER R., SOBEL M., "Correspondences between parts of shapes with particle filters", *IEEE Computer Vision and Pattern Recognition*, pp. 1–8, 2008.

[LAM 09] LAM S., CANE B.M., ALLEN R., "Automated tracking in digitized video fluoroscopy sequences for spine kinematic analysis", *Image and Vision Computing*, vol. 27, no. 10, pp. 1555–1571, 2009.

[LAN 06] LANZ O., "Approximate Bayesian multibody tracking", *IEEE Transactions on Pattern Analysis and Machine Intelligence*, vol. 28, no. 9, pp. 1436–1449, 2006.

[LAO 09] LAO Y., ZHU J., ZHENG Y., "Sequential particle generation for visual tracking", *IEEE Transactions on Circuits and Systems for Video Technology*, vol. 19, no. 9, pp. 1365–1378, 2009.

[LEI 06] LEI Y., DING X., WANG S., "AdaBoost tracker embedded in adaptive particle filtering", *IEEE Conference on Computer Vision and Pattern Recognition*, pp. 939–943, 2006.

[LEI 08] LEI Y., DING X., WANG S., "Visual tracker using sequential Bayesian learning: discriminative, generative, and hybrid", *IEEE Transactions on Systems, Man, and Cybernetics, Part B*, vol. 38, no. 6, pp. 1578–1591, 2008.

[LEV 44] LEVENBERG K., "A method for the solution of certain non-linear problems in least squares", *The Quarterly of Applied Mathematics*, vol. 2, pp. 164–168, 1944.

[LI 03] LI P., ZHANG T., PECE A., "Visual contour tracking based on particle filters", *Image and Vision Computing*, vol. 21, no. 1, pp. 111–123, 2003.

[LI 04a] LI P., CHAUMETTE F., "Image cues fusion for object tracking based on particle filter", *Workshop on Articulated Motion and Deformable Objects*, pp. 99–107, 2004.

[LI 04b] LI P., ZHANG T., MA B., "Unscented Kalman filter for visual curve tracking", *Image and Vision Computing*, vol. 22, pp. 157–164, 2004.

[LI 11] LI S., TAN M., TSANG I., *et al.*, "A hybrid PSO-BFGS strategy for global optimization of multimodal functions", *IEEE Transactions on Systems, Man, and Cybernetics – Part B*, vol. 41, no. 4, pp. 1003–1014, 2011.

[LIA 09] LIANGCHENG C., HUIZHONG Y., "A learning method of support vector machine based on particle filters", *International Conference on Natural Computation*, pp. 431–435, 2009.

[LIC 04] LICHTENAUER J., REINDERS M., HENDRIKS E., "Influence of the observation likelihood function on particle filtering performance in tracking applications", *IEEE International Conference on Automatic Face and Gesture Recognition*, pp. 767–772, 2004.

[LIU 98] LIU J., CHEN R., "Sequential Monte-Carlo methods for dynamic systems", *Journal of the American Statistical Association*, vol. 93, pp. 1032–1044, 1998.

[LIU 09] LIU J., TONG X., LI W., *et al.*, "Automatic player detection, labeling and tracking in broadcast soccer video", *Pattern Recognition Letters*, vol. 30, no. 2, pp. 103–113, 2009.

[LIU 12] LIU J., YANG Y., SALEEMI I., *et al.*, "Learning semantic features for action recognition via diffusion maps", *Computer Vision and Image Understanding*, vol. 116, no. 3, pp. 361–377, 2012.

[LIV 12] LIVNE M., SIGNAL L., TROJE N., *et al.*, "Human attributes from 3D pose tracking", *Computer Vision and Image Understanding*, vol. 116, pp. 648–660, 2012.

[LU 09] LU W., OKUMA K., LITTLE J., "Tracking and recognizing actions of multiple hockey players using the boosted particle filter", *Image and Vision Computing*, vol. 27, nos. 1–2, pp. 189–205, 2009.

[LUC 81] LUCAS B., KANADE T., "An iterative image registration technique with an application to stereo vision", *International Joint Conference on Artificial Intelligence*, pp. 674–679, 1981.

[LUO 12] LUO W., LI H., LIU G., *et al.*, "Global salient information maximization for saliency detection", *Signal Processing: Image Communication*, vol. 27, no. 3, pp. 238–248, 2012.

[MAC 98] MACKAY D.J.C., "Introduction to Monte-Carlo methods", *NATO Advanced Study Institute on Learning in graphical models*, pp. 175–204, 1998.

[MAC 99a] MACCORMICK J., BLAKE A., "A probabilistic exclusion principle for tracking multiple objects", *International Conference of Computer Vision*, pp. 572–578, 1999.

[MAC 99b] MACCORMICK J., BLAKE A., "A probabilistic exclusion principle for tracking multiple objects", *International Conference on Computer Vision*, pp. 572-587, 1999.

[MAC 00a] MACCORMICK J., Probabilistic modeling and stochastic algorithms for visual localisation and tracking, PhD thesis, Oxford University, 2000.

[MAC 00b] MACCORMICK J., ISARD M., "Partitioned sampling, articulated objects, and interface-quality hand tracking", *European Conference on Computer Vision*, pp. 3–19, 2000.

[MAG 05a] MAGGIO E., CAVALLARO A., "Hybrid Particle Filter and Mean Shift tracker with adaptive transition model", *IEEE International Conference on Acoustics, Speech, and Signal Processing*, pp. 221–224, 2005.

[MAG 05b] MAGGIO E., SMERALDI F., CAVALLARO A., "Combining colour and orientation for adaptive particle filter-based tracking", *British Machine Vision Conference*, pp. 659–668, 2005.

[MAG 07] MAGGIO E., SMERLADI F., CAVALLARO A., "Adaptive multifeature tracking in a particle filtering framework", *IEEE Transactions on Circuits and Systems for Video Technology*, vol. 17, no. 10, pp. 1348–1359, 2007.

[MAG 09] MAGGIO E., CAVALLARO A., "Accurate appearance-based Bayesian tracking for maneuvering targets", *Computer Vision and Image Understanding*, vol. 113, no. 4, pp. 544–555, 2009.

[MAR 63] MARQUARDT D., "An algorithm for least-squares estimation of nonlinear parameters", *Journal of the Society for Industrial and Applied Mathematics*, vol. 11, no. 2, pp. 431–441, 1963.

[MAS 01] MASKELL S., GORDON N., "A Tutorial on Particle Filters for On-line Nonlinear/Non-Gaussian Bayesian Tracking", *IEEE Transactions on Signal Processing*, vol. 50, no. 2, pp. 174–188, 2001.

[MEI 09] MEI X., LING H., "Robust visual tracking using L1 minimization", *International Conference on Computer Vision*, pp. 1436–1443, 2009.

[MER 00] VAN DER MERWE R., DOUCET A., DE FREITAS N., *et al.*, "The Unscented Particle Filter", *Advances in Neural Information Processing Systems*, pp. 584-590, 2000.

[MIN 86] MINOUX M., *Mathematical Programming: Theory and Algorithms*, Wiley-Interscience series in discrete mathematics and optimization, Wiley, 1986.

[MIN 10] MINVIELLE P., DOUCET A., MARRS A., *et al.*, "A Bayesian approach to joint tracking and identification of Geometric Shapes in Video Sequences", *Image and Vision Computing*, vol. 28, no. 1, pp. 111–123, 2010.

[MOR 95] MORAL P.D., NOYER J., RIGAL G., *et al.*, "Résolution particulaire et traitement non-linéaire du signal : applications radar/sonar", *Traitement du Signal*, vol. 12, no. 4, pp. 287–301, 1995.

[MOR 08] MORENO-NOGUER F., SANFELIU A., SAMARAS D., "Dependent multiple cue integration for robust tracking", *IEEE Transactions on Pattern Analysis and Machine Intelligence*, vol. 30, no. 4, pp. 670–685, 2008.

[MUN 08a] MUÑOZ-SALINAS R., AGUIRRE E., GARCIASILVENTE M., *et al.*, "A multiple object tracking approach that combines colour and depth information using a confidence measure", *Pattern Recognition Letters*, vol. 29, no. 10, pp. 1504–1514, 2008.

[MUN 08b] MUÑOZ-SALINAS R., GARCIASILVENTE M., MEDINACARNICER R., "Adaptive multi-modal stereo people tracking without background modeling", *Journal of Visual Communication and Image Representation*, vol. 19, no. 2, pp. 75–91, 2008.

[MUÑ 09] MUÑOZ-SALINAS R., MEDINA-CARNICER R., MADRID-CUEVAS F.J. *et al.*, "People detection and tracking with multiple stereo cameras using particle filters", *Journal of Visual Communication and Image Representation*, vol. 20, no. 5, pp. 339–350, 2009.

[MUÑ 10] MUÑOZ-SALINAS R., MEDINA-CARNICER R., MADRID-CUEVAS F., *et al.*, "Particle filtering with multiple and heterogeneous cameras", *Pattern Recognition*, vol. 43, no. 7, pp. 2390–2405, 2010.

[MUR 02] MURPHY K., Dynamic Bayesian networks: representation, inference and learning, PhD thesis, UC Berkeley, Computer Science Division, 2002.

[NEJ 08] NEJHUM S.M.S., HO J., YANG M.-H., "Visual tracking with histograms and articulating Blocks", *IEEE Conference Computer Vision and Pattern Recognition*, pp. 1–8, 2008.

[NIC 08] NICKEL K., STIEFELHAGEN R., "Dynamic integration of generalized cues for person tracking", *European Conference on Computer Vision*, pp. 514–526, 2008.

[NIE 04] NIETHAMMER M., TANNENBAUM A., "Dynamic geodesic snakes for visual tracking", *IEEE Conference on Computer Vision and Pattern Recognition*, vol. 1, pp. 660–667, 2004.

[NOR 00] NORTH B., BLAKE A., ISARD M., "Learning and classification of complex dynamics", *IEEE Transactions on Pattern Analysis and Machine Intelligence*, vol. 22, no. 8, pp. 781–796, 2000.

[NUM 02] NUMMIARO K., KOLLER-MEIER E., GOOL L.V., "Object tracking with an adaptive color-based particle filter", *Symposium for Pattern Recognition of the DAGM*, pp. 353–360, 2002.

[NUM 03a] NUMMIARO K., KOLLER-MEIER E., GOOL L.V., "An adaptive color-based particle filter", *Image and Vision Computing*, vol. 21, no. 1, pp. 99–110, 2003.

[NUM 03b] NUMMIARO K., KOLLER-MEIER E., GOOL L.V., "Color features for tracking non-rigid objects", *Chinese Journal of Automation, Special Issue on Visual Surveillance*, vol. 29, pp. 345–355, 2003.

[OIK 11] OIKONOMIDIS I., KYRIAZIS N., "Full DOF tracking of a hand interacting with an object by modeling occlusions and physical constraints", *IEEE International Conference on Computer Vision*, pp. 2088–2095, 2011.

[OIK 12] OIKONOMIDIS I., KYRIAZIS N., ARGYROS A., "Tracking the articulated motion of two strongly interacting hands", *IEEE Conference on Computer Vision and Pattern Recognition*, pp. 639–646, 2012.

[ORT 02] ORTON M., FITZERALD W., "A Bayesian approach to tracking multiple targets using sensor arrays and particle filters", *IEEE Transactions Pattern Analysis and Machine Intelligence*, vol. 50, no. 2, pp. 216–223, 2002.

[OSH 88] OSHER S.J., SETHIAN J.A., "Fronts propagation with curvature dependent speed: algorithms based on Hamilton-Jacobi formulations", *Journal of Computation Physics*, vol. 79, pp. 12–49, 1988.

[PAL 08] PALAIO H., BATISTA J., "A region covariance embedded in a particle filter for multi-objects tracking", *International Workshop on Visual Surveillance*, pp. 1–8, 2008.

[PAN 08a] PANTRIGO J., SANCHEZ A., MONTEMAYOR A., "Multi-dimensional visual tracking using scatter search particle filter", *Pattern Recognition Letters*, vol. 29, no. 8, pp. 1160–1174, 2008.

[PAN 08b] PANTRIGO J.J., SÁNCHEZ A., MONTEMAYOR A.S., "Multi-dimensional visual tracking using scatter search particle filter", *Pattern Recognition*, vol. 29, pp. 1160–1174, 2008.

[PAR 00] PARAGIOS N., DERICHE R., "Geodesic active contours and level sets for the detection and tracking of moving objects", *Transactions on Pattern Analysis and Machine Intelligence*, vol. 22, pp. 266–280, 2000.

[PAT 10] PATRAS I., HANCOCK E., "Coupled prediction classification for robust visual tracking", *IEEE Transactions on Pattern Analysis and Machine Intelligence*, vol. 32, no. 9, pp. 1553–1567, 2010.

[PEA 88] PEARL J., *Probabilistic Reasoning in Intelligent Systems: Networks of Plausible Inference*, Morgan Kaufman Publishers, 1988.

[PEC 06] PECE A., "Contour tracking based on marginalized likelihood ratios", *Image and Vision Computing*, vol. 24, no. 3, pp. 301–317, 2006.

[PER 07] PERREAULT S., HEBERT P., "Median filtering in constant time", *IEEE Transactions on Image Processing*, vol. 16, no. 9, pp. 2389–2394, 2007.

[PER 08] PERNKOPF F., "Tracking of multiple targets using online learning for reference model adaptation", *IEEE Transactions on Systems, Man, and Cybernetics, Part B*, vol. 38, no. 6, pp. 1465–1475, 2008.

[PET 99a] PETERFREUND N., "Robust tracking of position and velocity with Kalman snakes", *IEEE Transactions on Pattern Analysis and Machine Intelligence*, vol. 21, pp. 564–569, 1999.

[PET 99b] PETERFREUND N., "The velocity snake: deformable contour for tracking in spatio-velocity space", *Computer Vision and Image Understanding*, vol. 73, pp. 346–356, 1999.

[PÉT 09] PÉTERI R., SILER O., Object tracking using joint visible and thermal infrared video sequences, Report, MIA Laboratory, 2009.

[PHA 01a] PHAM B., ZHANG J., "*A fuzzy shape specification system to support design for aesthetics*", Chapter *Soft Computing in Measurement and Information*, pp. 204–252, Springer, 2001.

[PHA 01b] PHAM B., "Representation of fuzzy shapes", *International Workshop on Visual Form*, pp. 239–248, 2001.

[PIT 99] PITT M., SHEPARD N., "Filtering via simulation: auxiliary particle filter", *Journal of the American Statistical Association*, vol. 94, pp. 590–599, 1999.

[PLA 01] PLANKERS R., FUA P., "Articulated soft objects for video-based body modeling", *IEEE International Conference on Computer Vision*, pp. 394–401, 2001.

[PON 11] PONCE V., GORGA M., BAR X., "Human behavior analysis from video data using bag-of-gestures", *International Joint Conference on Artificial Intelligence*, pp. 2836–2837, 2011.

[POR 05] PORIKLI F., "Integral histogram: a fast way to extract histograms in Cartesian spaces", *IEEE Conference on Computer Vision and Pattern Recognition*, pp. 829–836, 2005.

[PRE 07] PRESS W., TEUKOLSKY S., VETTERLING W., *Numerical Recipes: The Art of Scientific Programming,* 3rd ed., Cambridge University Press, 2007.

[PÉR 02] PÉREZ P., HUE C., VERMAAK J., "Color-based probabilistic tracking", *European Conference on Computer Vision*, pp. 661–675, 2002.

[PÉR 04] PÉREZ P., VERMAAK J., BLAKE A., "Data fusion for visual tracking with particles", *Proceedings of the IEEE*, vol. 92, no. 3, pp. 495–513, 2004.

[QU 07] QU W., SCHONFELD D., "Real-time decentralized articulated motion analysis and object tracking from videos", *IEEE Transactions on Image Processing*, vol. 16, no. 8, pp. 2129–2138, 2007.

[RAS 01] RASMUSSEN C., HAGER G., "Probabilistic data association methods for tracking complex visual objects", *IEEE Transactions on Pattern Analysis and Machine Intelligence*, vol. 23, no. 6, pp. 560–576, 2001.

[RAT 05] RATHI Y., VASWANI N., TANNENBAUM A., "Particle filtering for geometric active contours with application to tracking moving and deforming objects", *IEEE Conference on Computer Vision and Pattern Recognition*, pp. 2–9, 2005.

[RAT 07a] RATHI Y., VASWANI N., TANNENBAUM A., "A generic framework for tracking using particle filter with dynamic shape prior", *IEEE Transactions on Image Processing*, vol. 16, no. 5, pp. 1370–1382, 2007.

[RAT 07b] RATHI Y., VASWANI N., TANNENBAUM A., "Tracking deforming objects using particle filtering for geometric active contours", *IEEE Transactions on Pattern Analysis And Machine Intelligence*, vol. 29, no. 8, pp. 1470–1475, 2007.

[REI 79] REID D., "An algorithm for tracking multiple targets", *IEEE Transactions on Automation and Control*, vol. 24, no. 6, pp. 1016–1034, 1979.

[MAR 11] MARTINEZ DEL RINCON J., ORRITE C., MEDRANO C., "Rao-Blackwellised particle filter for color-based tracking", *Pattern Recognition Letters*, vol. 32, no. 2, pp. 210–220, 2011.

[ROB 05] ROBERT C.P., CASELLA G., *Monte-Carlo Statistical Methods*, Springer-Verlag, 2005.

[ROM 05] ROMDHANI S., VETTER T., "Estimating 3D shape and texture using pixel intensity, edges, specular highlights, texture constraints and a prior", *IEEE Conference on Computer Vision and Pattern Recognition*, pp. 986–993, 2005.

[RON 96] RONG L., BAR-SHALOM Y., "Tracking in clutter with nearest neighbor filter: analysis and performance", *IEEE transactions on aerospace and electronic systems*, vol. 32, no. 3, pp. 995–1010, 1996.

[ROO 11] VAN ROOTSELER R., SPREEUWERS L., VELDHUIS R., "Application of 3D morphable models to faces in video images", *Symposium on Information Theory in the Benelux*, pp. 34–41, 2011.

[ROS 08] ROSE C., SABOUNE J., CHARPILLET F., "Reducing particle filtering complexity for 3D motion capture using dynamic Bayesian networks", *Conference on Artificial Intelligence*, pp. 1396–1401, 2008.

[ROU 10] ROUGIER C., MEUNIER J., "3D Head trajectory using a single camera", *International Conference on Signal and Information Processing*, pp. 505–512, 2010.

[RUB 00] RUBNER Y., TOMASI C., GUIBAS L., "The Earth Mover's Distance as a metric for image retrieval", *International Journal on Computer Vision*, vol. 40, no. 2, pp. 99–121, 2000.

[RUD 12] RUDOY D., ZELNIK-MANOR L., "Viewpoint selection for human actions", *International Journal of Computer Vision*, vol. 97, no. 3, pp. 243–254, 2012.

[RUI 01] RUI Y., CHEN Y., "Better proposal distributions: object tracking using unscented particle filter", *IEEE Conference on Computer Vision and Pattern recognition*, pp. 786–793, 2001.

[SAG 12] SAGHAFI B., RAJAN D., "Human action recognition using Pose-based discriminant embedding", *Signal Processing: Image Communication*, vol. 27, no. 1, pp. 96–111, 2012.

[SÁN 05a] SÁNCHEZ A., PANTRIGO J., GIANIKELLIS K., "Combining particle filter and population-based metaheuristics for visual articulated motion tracking", *Electronic Letters on Computer Vision and Image Analysis*, vol. 5, no. 3, pp. 68–83, 2005.

[SAT 04] SATOH Y., OKATANI T., DEGUCHI K., "A color-based probabilistic tracking by using graphical models", *International Conference on Intelligent Robots and Systems*, pp. 631–636, 2004.

[SAV 12] SAVRAN A., SANKUR B., BILGE M., "Comparative evaluation of 3D vs. 2D modality for automatic detection of facial action units", *Pattern Recognition*, vol. 45, no. 2, pp. 767–782, 2012.

[SCH 01] SCHULZ D., BURGARD W., FOX D., "Tracking multiple moving targets with a mobile robot using particle filters and statistical data association", *IEEE International Conference on Robotics and Automation*, pp. 1665–1670, 2001.

[SCH 03] SCHULZ D., BURGARD W., FOX D., "People tracking with a mobile robots using sample-based joint probabilistic data association filters", *International Journal of Robotics Research*, vol. 22, no. 2, pp. 99–116, 2003.

[SCH 06] SCHIFF J., GOLDBERG K., "Automated intruder tracking using particle filtering and a network of binary motion sensors", *IEEE International Conference on Automation Science and Engineering*, pp. 580–587, 2006.

[SCH 07] SCHINDLER G., DELLAERT F., "A Rao-Blackwellized parts-constellation tracker, ", vol. 4358, Chapter *Dynamical Vision,* LNCS, pp. 178–189, 2007.

[SHA 48] SHANNON C., "A mathematical theory of communication", *Bell System Technical Journal*, vol. 27, pp. 379–423, 1948.

[SHE 03] SHEN C., VAN DEN HENGEL A., DICK A., "Probabilistic multiple cue integration for particle filter based tracking", *International Conference of Digital Image Computing: Techniques and Applications*, pp. 399–408, 2003.

[SHE 06] SHEN H., NELSON G., KENNEDY S.,"Automatic tracking of biological cells and compartments using particle filters and active contours", *Chemometrics and Intelligent Laboratory Systems*, vol. 82, nos. 1–2, pp. 276–282, 2006.

[SHI 94] SHI J., TOMASI C., "Good features to track", *IEEE Conference on Computer Vision and Pattern Recognition*, pp. 593–600, 1994.

[SHI 11] SHITRIT H., BERCLAZ J., FLEURET F., "Tracking multiple people under global appearance constraints", *IEEE International Conference on Computer Vision*, pp. 137–144, 2011.

[SHO 11] SHOTTON J., FITZGIBBON A., COOK M., "Real-time human pose recognition in parts from single depth images", *IEEE Conference on Computer Vision and Pattern Recognition*, pp. 1297–1304, 2011.

[SIG 03] SIGAL L., ISARD M., SIGELMAN B., "Attractive people: assembling loose-limbed models using non-parametric belief propagation", *Conference on Neural Information Processing Systems*, pp. 1539–1546, 2003.

[SIG 04] SIGAL L., BHATIA S., ROTH S., "Tracking loose-limbed people", *IEEE Conference on Computer Vision and Pattern Recognition*, pp. 421–428, 2004.

[SIG 10a] SIGAL L., BALAN A., "Humaneva: synchronized video and motion capture dataset and baseline algorithm for evaluation of articulated human motion", *International Journal of Computer Vision*, vol. 1–2, pp. 4–27, 2010.

[SIG 10b] SIGAL L., FLEET D., TROJE N., "Human attributes from 3D pose tracking", *International Conference on Computer Vision*, pp. 243–257, 2010.

[SIN 74] SINGER R., SEA R., HOUSEWRIGHT K., "Derivation and evaluation of improved tracking filters for use in dense multitarget environments", *IEEE Transactions on Information Theory*, vol. 20, no. 4, pp. 423–432, 1974.

[SIV 09] SIVIC J., "Effcient visual search of videos cast as text retrieval", *IEEE Transactions on Pattern Analysis and Machine Intelligence*, vol. 31, no. 4, pp. 591–605, 2009.

[SIZ 08] SIZINTSEV M., DERPANIS K., HOGUE A., "Histogram-based search: a comparative study", *IEEE Conference on Computer Vision and Pattern Recognition*, pp. 1–8, 2008.

[SMA 07] SMAL I., NIESSEN W., MEIJERING E., "Particle filtering for multiple object tracking in molecular cell biology", *IEEE Nonlinear Statistical Signal Processing Workshop*, pp. 129–132, 2007.

[SMA 08] SMAL I., MEIJERING E., DRAEGESTEIN K., "Multiple object tracking in molecular bioimaging by Rao-Blackwellized marginal particle filtering", *Medical Image Analysis*, vol. 12, no. 6, pp. 764–777, 2008.

[SMI 04] SMITH K., GATICA-PEREZ D., "Order matters: a distributed sampling method for multi-object tracking", *British Machine Vision Conference*, pp. 25–32, 2004.

[SNO 09] SNOEK J., HOEY J., STEWART L.,"Automated detection of unusual events on stairs", *Image and Vision Computing*, vol. 27, nos. 1–2, pp. 153–166, 2009.

[SOF 10] SOFKA M., ZHANG J., ZHOU S., "Multiple object detection by sequential Monte-Carlo and hierarchical detection network", *Computer Vision and Pattern Recognition*, pp. 1735–1742, 2010.

[SON 05] SONG T.L., LEE D.G., RYU J., "A probabilistic nearest neighbor filter algorithm for tracking in a clutter environment", *Signal Processing*, vol. 85, no. 10, pp. 2044–2053, 2005.

[SPE 03] SPENGLER M., SCHIELE B., "Towards robust multi-cue integration for visual tracking", *Machine Vision and Applications*, vol. 14, no. 1, pp. 50–58, 2003.

[SUD 04] SUDDERTH E., MANDEL M., FREEMAN W., "Visual hand tracking using nonparametric belief propagation", *IEEE Conference on Computer Vision and Pattern Recognition Workshop*, pp. 189–197, 2004.

[SUD 10] SUDDERTH E.B., IHLER A., ISARD M., "Nonparametric belief propagation", *Communications of ACM*, vol. 53, pp. 95–103, 2010.

[SUL 01] SULLIVAN J., RITTSCHER J., "Guiding random particles by deterministic search", *International Conference on Computer Vision*, pp. 323–330, 2001.

[SZE 10] SZELISKI R. (ed.), *Computer Vision: Algorithms and Applications*, Springer, 2010.

[TAM 06] TAMMINEN T., LAMPINEN J., "Sequential Monte-Carlo for Bayesian matching of objects with occlusions", *IEEE Transactions on Pattern Analysis And Machine Intelligence*, vol. 28, no. 6, pp. 930–941, 2006.

[TAN 79] TANG G., YANG G., HUANG T., "A fast two-dimensional median filtering algorithm", *IEEE Transactions on Acoustics, Speech and Signal Processing*, vol. 27, no. 1, pp. 13–18, 1979.

[TAN 96] TANIZAKI H., MARIANO R.S., "Nonlinear filters based on Taylor series expansions", *Communications in Statistics – Theory and Methods*, vol. 25, pp. 1261–1282, 1996.

[TAN 09] TANÁCS A., DOMOKOS C., SLADOJE N., "Recovering affine deformations of fuzzy shapes", *Scandinavian Conference on Image Analysis*, pp. 735–744, 2009.

[TER 92] TERZOPOULOS D., SZELISKI R., "Tracking with Kalman snakes,", Chapter *Active Vision*, MIT Press, pp. 3–20, 1992.

[THI 12] THI T., CHENG L., ZHANG J., "Integrating local action elements for action analysis", *Computer Vision and Image Understanding*, vol. 116, no. 3, pp. 378–395, 2012.

[TIA 01] TIAN Y., KANADE T., COHN J., "Recognizing action units for facial expression analysis", *IEEE Transactions on Pattern Analysis and Machine Intelligence*, vol. 23, no. 2, pp. 976–115, 2001.

[TOR 01] TORMA P., SZEPESVÁRI C., "LS-N-IPS: an improvement of particle filters by means of local search", *Non-Linear Control Systems*, pp. 715–719, 2001.

[TOR 04] TORMA P., SZEPESVÁRI C., "Enhancing particle filters using local likelihood sampling", *European Conference on Computer Vision*, pp. 16–27, 2004.

[TRI 01] TRIESCH J., MALSBURG C.V.D., "Democratic integration: self-organized integration of adaptive cues", *Neural Computation*, vol. 13, pp. 2049–2074, 2001.

[VAN 11] VAN ROOTSELER R.T.A., SPREEUWERS L.J., VELDHUIS R.N.J., "Application of 3D morphable models to faces in video image", *Symposium on Information Theory in the Benelux*, pp. 34–41, 2011.

[VER 05a] VERMAAK J., GODSILL S., PÉREZ P., "Monte-Carlo filtering for multi target tracking and data association", *IEEE Transactions on Aerospace and Electronic Systems*, vol. 41, no. 1, pp. 309–332, 2005.

[VER 05b] VERMAAK J., IKOMA N., GODSILL S., "Sequential Monte-Carlo framework for extended object tracking", *IEEE Proceedings on Radar, Sonar and Navigation*, vol. 152, pp. 353–363, 2005.

[VIK 12] VIKRAM T., TSCHEREPANOW M., WREDE B., "A saliency map based on sampling an image into random rectangular regions of interest", *Pattern Recognition*, vol. 45, pp. 3114–3124, 2012.

[VON 08] VONDRAK M., SIGAL L., JENKINS O., "Physical simulation for probabilistic motion tracking", *IEEE International Conference on Computer Vision and Pattern Recognition*, pp. 1–8, 2008.

[WAN 05] WANG P., JI Q., "Multi-view face tracking with factorial and switching HMM", *IEEE Workshops on Application of Computer Vision*, pp. 401–406, 2005.

[WAN 06] WANG Q., XIE L., LIU J., "Enhancing particle swarm optimization based particle filter tracker", *International Conference on Intelligent computing: Part II*, pp. 1216–1221, 2006.

[WAN 07] WANG H., SUTER D., SCHINDLER K., "Adaptive object tracking based on an effective appearance filter", *IEEE Transactions on Pattern Analysis and Machine Intelligence*, vol. 29, no. 9, pp. 1661–1667, 2007.

[WAN 09] WANG Z., YANG X., XU Y., "CamShift guided particle filter for visual tracking", *Pattern Recognition Letters*, vol. 30, no. 4, pp. 407–413, 2009.

[WAN 11a] WANG C., SHEN S., LIU Y., "A fast approach to deformable surface 3D tracking", *Pattern Recognition*, vol. 44, no. 12, pp. 2915–2925, 2011.

[WAN 11b] WANG Z., "Time varying dynamic bayesian network for nonstationary events modeling and online inference", *IEEE Transactions on Signal Processing*, vol. 59, no. 4, pp. 1553–1568, 2011.

[WEI 06] VAN DE WEIJER J., GEVERS T., BAGDANOV A., "Boosting color saliency in image feature detection", *IEEE Transactions on Pattern Analysis and Machine Intelligence*, vol. 28, no. 1, pp. 150–156, 2006.

[WID 10a] WIDYNSKI N., Intégration d'informations spatiales floues dans un filtre particulaire pour le suivi mono- et multi-objets dans des séquences d'images 2D, PhD thesis, Télécom ParisTech, 2010.

[WID 10b] WIDYNSKI N., DUBUISSON S., BLOCH I., "Introducing fuzzy spatial constraints in a ranked partitioned sampling for multi-object tracking", *International Symposium on Visual Computing*, pp. 393–404, 2010.

[WID 11a] WIDYNSKI N., ALDEA E., DUBUISSON S., "Object tracking based on particle filtering with multiple appearance models", *International Conference on Computer Vision Theory and Applications*, pp. 604–609, 2011.

[WID 11b] WIDYNSKI N., DUBUISSON S., BLOCH I., "Integration of fuzzy spatial information in tracking based on particle filtering", *IEEE Transactions on Systems, Man and Cybernetics SMC-B*, vol. 41, no. 3, pp. 635–649, 2011.

[WID 12] WIDYNSKI N., DUBUISSON S., BLOCH I., "Fuzzy spatial constraints and ranked partitioned sampling approach for multiple object tracking", *Computer Vision and Image Understanding*, vol. 116, no. 10, pp. 267–282, 2012.

[WOO 07] WOODLEY T., STENGER B., CIPOLLA R., "Tracking using online feature selection and a local generative model", *British Machine Vision Conference*, pp. 1–10, 2007.

[WU 01] WU Y., HUANG T.S., "A co-inference approach to robust visual tracking", *International Conference on Computer Vision*, pp. 26–33, 2001.

[WU 08a] WU H., SUN F., LIU H., "Fuzzy particle filtering for uncertain systems", *IEEE Transactions on Fuzzy Systems*, vol. 16, no. 5, pp. 1114–1129, 2008.

[WU 08b] WU Y., TONG X., ZHANG Y., "Boosted interactively distributed particle filter for automatic multi-object tracking", *International Conference on Image Processing*, pp. 1844–1847, 2008.

[WU 12] WU S., WONG H., "Joint segmentation of collectively moving objects using a bag-of-words model and level set evolution", *Pattern Recognition*, vol. 45, pp. 3389–3401, 2012.

[WU 13] WU Y., LIM J., YANG M.-H., "Online object tracking: a benchmark", *IEEE Conference on Computer Vision and Pattern Recognition*, pp. 2411–2418, 2013.

[XIA 08] XIANG S., NIE F., SONG Y., "Contour graph based human tracking and action sequence recognition", *Pattern Recognition*, vol. 41, no. 12, pp. 3653–3664, 2008.

[YE 09] YE W., LIU H., SUN F., "Vehicle tracking based on co-learning particle filter", *IEEE International Conference on Intelligent Robots and Systems*, pp. 2979–2984, 2009.

[YED 05] YEDIDIA J., FREEMAN W., WEISS Y., "Constructing free-energy approximations and generalized belief propagation algorithms", *IEEE Transactions on Information Theory*, vol. 51, no. 7, pp. 2282–2312, 2005.

[YEZ 03] YEZZI A., SOATTO S., "Deformation: deforming motion, shape average and the joint registration and approximation of structures in images", *International Journal of Computer Vision*, vol. 53, pp. 153–167, 2003.

[YU 09] YU X., LIU J., TANG X., "Using Rao-Blackwellised particle filter track 3D arm motion based on hierarchical limb model", *WRI World Congress on Computer Science and Information Engineering*, vol. 2, pp. 85–89, 2009.

[ZAD 68] ZADEH L., "Probability measures of fuzzy events", *Journal of Mathematical Analysis and Applications*, vol. 23, no. 2, pp. 421–427, 1968.

[ZAD 75] ZADEH L., "The concept of a linguistic variable and its application to approximate reasoning – I", *Information Sciences*, vol. 8, no. 3, pp. 199–249, 1975.

[ZHA 10] ZHAO J., LI Z., "Particle filter based on Particle Swarm Optimization resampling for vision tracking", *Expert Systems with Applications*, vol. 37, pp. 8910–8914, 2010.

[ZHO 04] ZHOU S., CHELLAPPA R., MOGHADDAM B., "Appearance tracking using adaptive models in a particle filter", *Asian Conference on Computer Vision*, pp. 1–6, 2004.

[ZHO 12] ZHOU B., WANG X., TANG X., "Understanding collective crowd behaviors: learning a mixture model of dynamic pedestrian-agents", *IEEE Conference on Computer Vision and Pattern Recognition*, pp. 1–8, 2012.

[ZHU 10] ZHU Y., FUJIMURA K., "A Bayesian framework for human body pose tracking from depth image sequences", *Sensors*, vol. 10, no. 5, pp. 5280–5293, 2010.

Index

B, D, E

Bayesian
  filtering, 1, 2, 4
  networks, 26, 134, 155, 158
behavior analysis, 152
Bhattacharyya, 14, 16, 46–50, 60, 61, 86
data association, 53, 80–82, 87, 120
deformation, 20, 30, 50, 53–55, 76, 116
erratic movements, 8, 103, 104
event detection, 156

H, L, M

histogram, 30
likelihood function, 8, 12, 14, 15, 30, 75
movement detection, 50
multi-object tracking, 80, 82, 87, 94, 106, 119, 125, 127, 149
multiappearance, 69, 76
multimodal, 51, 59, 76, 150, 155, 156, 160

O, P, R

observation model, 10, 19, 60
optimization, 23, 24, 36, 79, 80, 115, 116, 121, 140, 156, 157
partition sampling, 125
resampling, 7, 16, 143

S, V

social interactions, 153, 159, 160
space decomposition, 140
spatial relations, 80, 96, 98, 99, 100, 106, 121, 158
state model, 8–11
swapping, 141
visual tracking, 8

Other titles from

iSTE

in

Digital Signal and Image Processing

2014

AUGER François
Signal Processing with Free Software: Practical Experiments

BLANCHET Gérard, CHARBIT Maurice
Digital Signal and Image Processing using MATLAB
Volume 1 – Fundamentals – 2nd edition

FANET Hervé
Medical Imaging Based on Magnetic Fields and Ultrasounds

MOUKADEM Ali, OULD Abdeslam Djaffar, DIETERLEN Alain
Time-Frequency Domain for Segmentation and Classification of Non-stationary Signals: The Stockwell Transform Applied on Bio-signals and Electric Signals

NDAGIJIMANA Fabien
Signal Integrity: From High Speed to Radiofrequency Applications

PINOLI Jean-Charles
Mathematical Foundations of Image Processing and Analysis – Volumes 1 and 2

TUPIN Florence, INGLADA Jordi, NICOLAS Jean-Marie
Remote Sensing Imagery

VLADEANU Calin, EL ASSAD Safwan
Nonlinear Digital Encoders for Data Communications

2013

GOVAERT Gérard, NADIF Mohamed
Co-Clustering

DAROLLES Serge, DUVAUT Patrick, JAY Emmanuelle
Multi-factor Models and Signal Processing Techniques: Application to Quantitative Finance

LUCAS Laurent, LOSCOS Céline, REMION Yannick
3D Video: From Capture to Diffusion

MOREAU Eric, ADALI Tulay
Blind Identification and Separation of Complex-valued Signals

PERRIN Vincent
MRI Techniques

WAGNER Kevin, DOROSLOVACKI Milos
Proportionate-type Normalized Least Mean Square Algorithms

FERNANDEZ Christine, MACAIRE Ludovic, ROBERT-INACIO Frédérique
Digital Color Imaging

FERNANDEZ Christine, MACAIRE Ludovic, ROBERT-INACIO Frédérique
Digital Color: Acquisition, Perception, Coding and Rendering

NAIT-ALI Amine, FOURNIER Régis
Signal and Image Processing for Biometrics

OUAHABI Abdeljalil
Signal and Image Multiresolution Analysis

2011

CASTANIÉ Francis
Digital Spectral Analysis: Parametric, Non-parametric and Advanced Methods

DESCOMBES Xavier
Stochastic Geometry for Image Analysis

FANET Hervé
Photon-based Medical Imagery

MOREAU Nicolas
Tools for Signal Compression

2010

NAJMAN Laurent, TALBOT Hugues
Mathematical Morphology

2009

BERTEIN Jean-Claude, CESCHI Roger
Discrete Stochastic Processes and Optimal Filtering / 2nd edition

CHANUSSOT Jocelyn *et al.*
Multivariate Image Processing

DHOME Michel
Visual Perception through Video Imagery

GOVAERT Gérard
Data Analysis

GRANGEAT Pierre
Tomography

MOHAMAD-DJAFARI Ali
Inverse Problems in Vision and 3D Tomography

SIARRY Patrick
Optimisation in Signal and Image Processing

2008

ABRY Patrice *et al.*
Scaling, Fractals and Wavelets

GARELLO René
Two-dimensional Signal Analysis

HLAWATSCH Franz *et al.*
Time-Frequency Analysis

IDIER Jérôme
Bayesian Approach to Inverse Problems

MAÎTRE Henri
Processing of Synthetic Aperture Radar (SAR) Images

MAÎTRE Henri
Image Processing

NAIT-ALI Amine, CAVARO-MENARD Christine
Compression of Biomedical Images and Signals

NAJIM Mohamed
Modeling, Estimation and Optimal Filtration in Signal Processing

QUINQUIS André
Digital Signal Processing Using Matlab

2007

BERTEIN Jean-Claude, CESCHI Roger
Discrete Stochastic Processes and Optimal Filtering

BLOCH Isabelle
Information Fusion in Signal and Image Processing

GLAVIEUX Alain
Channel Coding in Communication Networks

OPPENHEIM Georges *et al.*
Wavelets and their Applications

2006

CASTANIÉ Francis
Spectral Analysis

NAJIM Mohamed
Digital Filters Design for Signal and Image Processing

Printed and bound by CPI Group (UK) Ltd, Croydon, CR0 4YY

13/09/2026

14961078-0002